Lecture Notes in Biomathematics **97**

Managing Editor:

S.A. Levin

Vincenzo Capasso

Mathematical Structures of Epidemic Systems

Springer

Vincenzo Capasso
Dipartimento di Matematica
Università degli Studi di Milano
Via Saldini, 50
20133 MILANO, Italy
e-mail: vincenzo.capasso@unimi.it

ISBN 978-3-540-56526-0 e-ISBN 978-3-540-70514-7

Lecture Notes in Biomathematics ISSN 0341-633X

Library of Congress Catalog Number: 2008929585

Mathematics Subject Classification (2000): 92D30, 35K57, 34C12, 37C65, 34CXX, 35BXX

Corrected 2nd printing 2008

Cover design: WMXDesign GmbH, Heidelberg, Germany

Printed on acid-free paper

9 8 7 6 5 4 3 2 1

springer.com

A chi mi ha dedicato
tutti i suoi pensieri.

Foreword

The dynamics of infectious diseases represents one of the oldest and richest areas of mathematical biology. From the classical work of Hamer (1906) and Ross (1911) to the spate of more modern developments associated with Anderson and May, Dietz, Hethcote, Castillo-Chavez and others, the subject has grown dramatically both in volume and in importance. Given the pace of development, the subject has become more and more diffuse, and the need to provide a framework for organizing the diversity of mathematical approaches has become clear. Enzo Capasso, who has been a major contributor to the mathematical theory, has done that in the present volume, providing a system for organizing and analyzing a wide range of models, depending on the structure of the interaction matrix. The first class, the quasi-monotone or positive feedback systems, can be analyzed effectively through the use of comparison theorems, that is the theory of order-preserving dynamical systems; the second, the skew-symmetrizable systems, rely on Lyapunov methods. Capasso develops the general mathematical theory, and considers a broad range of examples that can be treated within one or the other framework. In so doing, he has provided the first steps towards the unification of the subject, and made an invaluable contribution to the Lecture Notes in Biomathematics.

Simon A. Levin

Princeton, January 1993

Author's Preface to Second Printing

In the Preface to the First Printing of this volume I wrote:

" ..[I] hope to find some reader who may appreciate the volume as a guided tour through the vast literature on the subject."

I am glad, after such a long time (about twenty years) to have discovered that my book received much more attention than expected.

I wish to thank Catriona Byrne, the Mathematical Editor of Springer-Heidelberg, who kindly insisted that the book be reprinted, thus making it available again after many requests that could be not satisfied, since the original printing was sold out.

I have taken the opportunity, in this second printing, to correct all detected misprints. I have also included reference data to papers in the bibliography that have meanwhile been published.

Vincenzo Capasso

Milan, May 2008

"Non con soverchie speranze ...,
né avendo nell'animo illusioni
spesso dannose, ma nemmeno con
indifferenza, deve essere accolto
ogni tentativo di sottoporre al calcolo
fatti di qualsiasi specie."
(Vito Volterra, 1901)

Author's Preface

It is now exactly twenty years since the first time I read the first edition of the now classic book by N.T.J. Bailey, The Mathematical Theory of Epidemics (Griffin, London, 1957). With my background in Theoretical Physics, I had been attracted by the possibility of analyzing with mathematical rigor an area of Science which deals with highly complex natural systems. Anyway, in the preface of his book, Bailey stated that the discipline was already old about fifty years, in the modern sense of the phrase, by dating the beginnings at the work by William Hamer (1906) and Ronald Ross (1911).

This monograph was started after a suggestion by Simon A. Levin, during an Oberwolfach workshop in 1984, to organize better my own ideas about the mathematical structures of epidemic systems, that I had been presenting in various papers and conferences. He had been very able to identify the "leit motiv" of my thoughts, that a professional mathematician can contribute in the growth of knowledge only if he is capable of building up a fair and correct interface between the core subject of a specific discipline and the most recent "tools" of Mathematics.

The scope of this monograph is then to make them available to a large audience, in a possibly accessible way, powerful techniques of modern Mathematics, without obscuring with "magic symbols" the intrinsic vitality of mathematical concepts and methods.

> "I non iniziati ai segreti del Calcolo e dell'Algebra si fanno talora l'illusione che i loro mezzi siano di natura diversa da quelli di cui il comune ragionamento dispone." (Volterra,1901).

Clearly I did not go much further than my wishful thinking, but still hope to find some reader who may appreciate the volume as a guided tour through the vast literature on the subject. I wish to specify that the list of references includes only the ones explicitly quoted in the text. I apologize for my ignorance of papers directly related with this monograph.

The contribution of Dr. R. Caselli is warmly acknowledged for all the numerical simulations and their graphical representation included in the monograph.

It is now time to thank Si for his encouragement and patience. Also for her very gentle patience I wish to thank Dr. C. Byrne (Mathematical Editor of Springer-Verlag) who has been waiting and supporting this project for such a long time.

I shall not forget to thank the Director and the staff of the Mathematical Centre at Oberwolfach for providing me, during a wonderful month in the summer of 1990, the right scientific environment for producing the core of this monograph.

Thanks are due to the numerous Colleagues who carefully read parts of the manuscript, and gave me relevant advice ; in particular I thank Edoardo Beretta, Carlos Castillo-Chavez, Andrea di Liddo, Herb Hethcote, Mimmo Iannelli, John Jacquez, Simon Levin, Stefano Paveri-Fontana, Andrea Pugliese, Carl Simon.

I also wish to thank S. Levin and coauthors for the use of Figures 3.1, 3.3 and Tables 3.1-3.5; J. Jacquez and coauthors for Figures 3.5, 3.6; H. Hethcote and coauthors for Table 3.6.

Finally I would like to thank my research advisor at the University of Maryland (College Park) Grace Yang, for the key role played in introducing me to this very challenging area of scientific research, and Jim Murray for making me familiar with reaction-diffusion systems.

Financial assistance is acknowledged by the National Research Council of Italy (CNR) through the National Group for Mathematical Physics (GNFM) and the Institute for Research in Applied Mathematics (IRMA).

Vincenzo Capasso

Milan, October 1992

Table of Contents

"... l' universo ... é scritto in lingua matematica,
e i caratteri sono triangoli, cerchi, ed altre figure
geometriche ...; senza questi
é un aggirarsi vanamente per un oscuro laberinto"
(Galileo Galilei, Saggiatore (VI, 232), 1623).

"All epidemiology, conceived as it is with the
variation of disease from time to time and
from place to place, must be considered
mathematically, however many variables are
implicated, if it is to be considered
scientifically at all"
(Sir Ronald Ross, 1911)

1. Introduction

The main scope of mathematical modelling in epidemiology is clearly stated in the second edition (1975) of Bailey's book [19]: "we need to develop models that will assist the decision-making process by helping to evaluate the consequences of choosing one of the alternative strategies available. Thus , mathematical models of the dynamics of a communicable disease can have a direct bearing on the choice of an immunization program, the optimal allocation of scarce resources, or the best combination of control or eradication techniques."

We may like to say with Okubo [177] that "A mathematical treatment is indispensable if the dynamics of ecosystems are to be analyzed and predicted quantitatively. The method is essentially the same as that used in such fields as classical and quantum mechanics, molecular biology, and biophysics... One must not be enamored of mathematical models; there is no mystique associated with them...physics and mathematics must be considered as tools rather than sources of knowledge, tools that are effective but nonetheless dangerous if misused".

Even though I consider mathematical reasoning much more than just a tool in scientific investigation, in this monograph I have pursued the main objective of providing a companion in the scientific process of building and analyzing mathematical models for communicable diseases.

As reported in the long, but still a sample, list of references, an enormous literature is available nowadays, dealing with modelling the dynamics of infectious diseases (during the final phase of preparation of this monograph a monumental volume has appeared due to Anderson and May [9] which is further encouraging in this direction). What I personally feel is that there is a concrete possibility of classifying most of the available models according to their mathematical structure.

In this respect two main classes may be identified. One of them, composed of the quasimonotone or positive feedback systems, has attracted various mathematicians in the last twenty years to build a mathematical theory of order preserving dynamical systems. In the other case, Lyapunov methods play a central role.

The Italian main precursor in the field of biomathematics, Vito Volterra, in his pioneering work on predator-prey systems, introduced a Lyapunov functional (the Volterra-Lyapunov potential) which has been the basis for a large amount of work on the generalized Lotka-Volterra systems. As shown in this monograph, these include a large class of epidemic systems, based on the "law of mass action". (The Volterra-Lyapunov potential has been recently given an information theoretic interpretation by Capasso and Forte in [52]).

A lot of attention has been attracted in the recent years to the mathematical modelling of HIV/AIDS infection, in order to predict the evolution of this modern "plague". Actually this poses highly challenging problems, which are essentially of modelling more than of analysis. Due to the long duration of the disease in each individual, and to the fast transportation means between different geographical areas of the world, and the increased communication among different social groups, coupling at different time and "space" scales cannot be ignored. Problems of coupling at very different scales pose big challenges to mathematical analysis and computation. A chapter has been devoted to HIV/AIDS infections as a specific case study; but, in the spirit of this monograph, only simplified "educational models" have been analyzed.

Only purely deterministic models are the subject of this monograph, even though I think that in order to fit real data, stochastic fluctuations cannot be ignored, especially in connection with biological systems. Furthermore the analysis of most stochastic models is based on the common tools of the mathematical theory of evolution equations (ODE's and PDE's), so that this may provide the necessary background for stochastic modelling as well.

Who knows ? This might be the first of two volumes...

For the biological interpretation of the models which are analyzed here we refer to the literature, while for an historical development of the subject we refer to Dietz and Schenzle [83].

We shall mainly be concerned with the so called "compartmental models".

Compartmental models are most suitable for microparasitic infections (typified by most viral and bacterial, and many protozoan, infections) [163]; the duration of infection is usually short, relative to the expected life span of the host.

In a compartmental model the total population (relevant to the epidemic process) is divided into a number (usually small) of discrete categories: susceptibles, infected but not yet infective (latent), infective, recovered and immune, without distinguishing different degrees of intensity of infection.

In contrast, for macroparasitic infections, such as helminthic infections, it is relevant to know the parasite burden borne by an individual host: there can be an important distinction between infection (having one or more parasites) and disease (having a parasite load large enough to produce illness). Consequently, mathematical models for host-macroparasitic associations need to deal with the full distribution of parasites among the host population [82].

We shall not analyze this case, for which we refer to the literature (see e.g. [82, 92, 173]).

A key problem in modelling the evolution dynamics of infectious diseases is the mathematical representation of the mechanism of transmission of the contagion. The concepts of "force of infection" and "field of forces of infection" (when dealing with structured populations) which were introduced in [48], will be the guideline of this presentation.

Suppose at first that the population in each compartment does not exhibit any structure (space location, age, etc.). The infection process (S to I) is driven by a force of infection ($f.i.$) due to the pathogen material produced by the infective population and available at time t

$$(f.i.)(t) = [g(I(\cdot))]\,(t) \tag{1.1}$$

which acts upon each individual in the susceptible class. Thus a typical rate of the infection process is given by the

$$(incidence\ \ rate)(t) = (f.i.)(t)\ S(t). \tag{1.2}$$

From this point of view, the "law of mass action" simply corresponds to choosing a linear dependence of $g(I)$ upon I [132]

$$(f.i.)(t) = k\,I(t). \tag{1.3}$$

Section 2 is devoted to epidemic models based on the "law of mass action". From a mathematical point of view the evolution of the epidemic is described (in the space and time homogeneous cases) by ODE 's which contain at most bilinear terms. The major "tool" in analyzing these systems is the "Volterra-Lyapunov potential".

In Section 3 the law of mass action model has been extended to include a nonlinear dependence

$$(f.i.)(t) = g(I(t))\ ; \tag{1.4}$$

particular cases are

$$g(I) = k\ I^p\ , \qquad p > 0 \tag{1.5}$$

$$g(I) = \frac{k\ I^p}{\alpha + \beta\ I^q}\ , \qquad p, q > 0\ . \tag{1.6}$$

The general model (1.1) for the force of infection may be extended to include a nonlinear dependence upon both I and S , as discussed in the recent modelling of AIDS epidemics.

When dealing with populations which exhibit some structure (identified here by a parameter z) either discrete (e.g. social groups) or continuous (e.g. space location, age, etc.), the target of the infection process is the specific "subgroup" z in the susceptible class, so that the force of infection has to be evaluated with reference to that specific subgroup. This induces the introduction of a classical concept in physics: the "field of forces of infection" $(f.i.)(z;t)$ such that the incidence rate at time t at the specific "location" z will be given by

$$(incidence\ \ rate)(z;t) = (f.i.)(z;t)\ s(z;t). \tag{1.7}$$

We may like to remark here that this concept is not very far from the mediaeval idea that infectious diseases were induced into a human being by a flow of bad air ("mal aria" in Italian).

Anyhow in quantum field theory any field of forces is due to an exchange of particles: in this case bacteria, viruses, etc., so that the corpuscular and the continuous concepts of field are conceptually unified.

It is of interest to identify the possible structures of the field of forces of infection which depend upon the specific mechanisms of transmission of the disease among different groups. This problem has been raised since the very first models when age and/or space dependence had to be taken into account.

Section 5 is devoted to systems with space structure.

When dealing with populations with space structure the relevant quantities are spatial densities, such as $s(z;t)$ and $i(z;t)$, the spatial densities of susceptibles and of infectives respectively, at a point z of the habitat Ω, and at time $t \geq 0$.

The corresponding total populations are given by

$$S(t) = \int_\Omega s(z;t)\ dz \tag{1.8}$$

$$I(t) = \int_\Omega i(z;t)\ dz \tag{1.9}$$

In the law of mass action model, if only local interactions are allowed, the field at point $z \in \Omega$ is given by

$$(f.i.)(z;t) = k(z)\ i(z;t). \tag{1.10}$$

On the other hand if we wish to take also distant interactions into account as proposed by D.G. Kendall [130], the field at point $z \in \Omega$ is given by (see Section 5.5)

$$(f.i.)(z;t) = \int_{\Omega} k(z,z')\ i(z';t)\ dz'\ . \tag{1.11}$$

When dealing with populations with an age structure (see Section 6) we interpret the parameter z as the age-parameter so that model (1.10) is a model with intracohort interactions while model (1.11) is a model with intercohort interactions.

Section 4 and consequently large parts of Section 5 are devoted to mathematical models of communicable diseases, which exhibit a cooperative (positive feedback) structure. The common feature for this class of models is the monotonicity (order preservation) of the dynamical systems associated with the epidemic models.

The non monotone case has been also considered by means of Lyapunov functionals and the LaSalle Invariance Principle (see Section 5.6).

The emergence of travelling waves in epidemic systems with spatial structure will not be discussed here. An elegant introduction to the subject has been provided by J.D. Murray [171].

Chapter 7 contains a brief presentation on the use of mathematical models in the definition of optimal control strategies and in the key problem of identification of parameters.

Appendices A and B (more technical in nature) have been added for the ease of non professional mathematicians who may then find this monograph self consistent as an introduction to the mathematical modelling of infectious diseases.

2. Linear models

2.1. One population models

We shall start considering the evolution of an epidemic in a closed host population of total size N. One of the most elementary compartmental models is the so called SIR model which was first due to Kermack-McKendrick [132] but is reproposed here in a rather simplified structure (see also [19] and [9]).

The total population is divided into three classes:

(S) the class of susceptibles, i.e. those individuals capable of contracting the disease and becoming themselves infectives;
(I) the class of infectives, i.e. those individuals capable of transmitting the disease to susceptibles;
(R) the class of removed individuals, i.e. those individuals which, having contracted the disease, have died or, if recovered, are permanently immune, or have been isolated, thus being unable to further transmit the disease.

A model based on these three compartments is generally called a SIR model. In order to write down a mathematical formulation for the dynamics of the epidemic process we introduce differential equations for the rates of transfer from one compartment to another:

$$\begin{aligned} \frac{dS}{dt} &= f_1(I,S,R) \\ \frac{dI}{dt} &= f_2(I,S,R) \\ \frac{dR}{dt} &= f_3(I,S,R) \end{aligned} \tag{2.1}$$

Typically a "law of mass action" [105, 222] has been assumed for the infection process: the transfer process from S to I. On the other hand the transfer from I to R is considered to be a pure exponential decay.

Thus the simplest choice for f_i , $i = 1,2,3$ has been the following:

$$\begin{aligned} f_1(I,S,R) &= -kIS \\ f_2(I,S,R) &= +kIS - \lambda I \\ f_3(I,S,R) &= +\lambda I \end{aligned} \tag{2.2}$$

with k and λ positive constants.

It is easily understood that in (2.2) it is assumed that when a susceptible is infected he immediately becomes infectious, i.e. there is no latent period.

If latency is allowed, an additional class (E) of latent individuals may be included (see Section 3).

2.1.1. SIR model with vital dynamics

In the above formulation the total population

$$N = S + I + R \tag{2.3}$$

is a constant, as can be seen by simply adding the three equations in (2.2).

The invariance of the total population can be maintained if we introduce an intrinsic vital dynamics of the individuals in the total population by means of a net mortality μN compensated by an equal birth input in the susceptible class.

In this case (2.2) are substituted by:

$$\begin{aligned} f_1(I,S,R) &= -kIS - \mu S + \mu N \\ f_2(I,S,R) &= +kIS - \lambda I - \mu I \\ f_3(I,S,R) &= \lambda I - \mu R \end{aligned} \tag{2.4}$$

In fact, it is easy to check that

$$N(t) = S(t) + I(t) + R(t) \tag{2.5}$$

is again constant in time.

We shall assume model (2.4) as a convenient point of departure for subsequent analysis, since it already contains the basic features of a general epidemic system, including the possibility of a nontrivial steady state as we shall see later.

System (2.1) together with (2.4) becomes,

$$\begin{cases} \dfrac{dS}{dt} = -kIS - \mu S + \mu N \\ \dfrac{dI}{dt} = kIS - \mu I - \lambda I \\ \dfrac{dR}{dt} = \lambda I - \mu R \end{cases} \tag{2.6}$$

for $t > 0$, which has to be subject to suitable initial conditions.

In this same class other models can be introduced. We shall list the most well known. From now on, when constant in time, the total population N will be assumed equal to 1, so that we refer to fractions of the total population. For a discussion about the related values of the parameters, refer to [118].

The SIR model with vital dynamics will then be rewritten as follows:

$$(2.6') \qquad \begin{cases} \dfrac{dS}{dt} = -kIS - \delta S + \delta \\ \dfrac{dI}{dt} = kIS - \gamma I - \delta I \\ \dfrac{dR}{dt} = \gamma I - \delta R \end{cases}$$

We may notice that the first two equations may be solved independently of the third one. Thus we shall be limiting ourselves to a two-dimensional system.

The same will be done in other cases without further advice.

2.1.2. SIRS model with temporary immunity [110]

This model derives from the SIR model with vital dynamics, but recovery gives only a temporary immunity

$$(2.7) \qquad \begin{cases} \dfrac{dS}{dt} = -kIS + \delta - \delta S + \alpha R \\ \dfrac{dI}{dt} = kIS - (\gamma + \delta)I \\ \dfrac{dR}{dt} = \gamma I - \alpha R \end{cases}$$

2.1.3. SIR model with carriers [110]

A carrier is an individual who carries and spreads the infectious disease, but has no clinical symptoms. If we assume that the number C of the carriers in the population is constant, we modify accordingly the SIR model with vital dynamics,

$$(2.8) \qquad \begin{cases} \dfrac{dS}{dt} = -k(I + C)S + \delta - \delta S \\ \dfrac{dI}{dt} = k(I + C)S - (\gamma + \delta)I \\ \dfrac{dR}{dt} = \gamma I - \delta R \end{cases}$$

2.1.4. The general structure of bilinear systems

According to a recent formulation due to Beretta and Capasso [28] all of the above models can be written in the general form:

$$\frac{dz}{dt} = diag(z)(e + Az) + c \tag{2.9}$$

where

$z \in \mathbb{R}^n$,	n	being the number of different compartments
$e \in \mathbb{R}^n$,		is a constant vector
$A = (a_{ij})_{i,j=1,\ldots,n}$		is a real constant matrix
$c \in \mathbb{R}^n$,		is a constant vector.

In the above examples we have in fact:

- SIR model with vital dynamics (model (2.6))

$$A = \begin{pmatrix} 0 & -k \\ k & 0 \end{pmatrix}; \quad e = \begin{pmatrix} -\delta \\ -(\delta+\gamma) \end{pmatrix}; \quad c = \begin{pmatrix} \delta \\ 0 \end{pmatrix} \tag{2.10}$$

- SIRS model with temporary immunity (model (2.7))

For our convenience, we change the variables (S, I) into $(\tilde{S}, I)$ such that $\tilde{S} = S + \frac{\alpha}{k}$.

Again, by taking into account that $S + R + I = 1$ (constant in time), we may ignore the equation for R.

Thus system (2.1) becomes:

$$(2.11) \qquad \begin{cases} \dfrac{d\tilde{S}}{dt} = -(\delta+\alpha)\tilde{S} - k\tilde{S}I + (\delta+\alpha)\left(1+\dfrac{\alpha}{k}\right) \\ \dfrac{dI}{dt} = -(\gamma+\delta+\alpha)I + k\tilde{S}I \end{cases}$$

so that

$$A = \begin{pmatrix} 0 & -k \\ k & 0 \end{pmatrix}; \quad e = \begin{pmatrix} -(\delta+\alpha) \\ -(\gamma+\delta+\alpha) \end{pmatrix}; \quad c = \begin{pmatrix} (\delta+\alpha)\left(1+\dfrac{\alpha}{k}\right) \\ 0 \end{pmatrix}$$

- SIR model with carriers (model 2.8)).

We change the variables (S, I) into $(S, \tilde{I})$, with $\tilde{I} = I + C$, so that system (2.8) becomes, ignoring the equation for R,

$$(2.12) \qquad \begin{cases} \dfrac{dS}{dt} = -\delta S - k\tilde{I}S + \delta \\ \dfrac{d\tilde{I}}{dt} = -(\gamma+\delta)\tilde{I} + k\tilde{I}S + (\gamma+\delta)C \end{cases}$$

Hence

$$A = \begin{pmatrix} 0 & -k \\ k & 0 \end{pmatrix}; \quad e = \begin{pmatrix} -\delta \\ -(\gamma+\delta) \end{pmatrix}; \quad c = \begin{pmatrix} \delta \\ (\gamma+\delta)C \end{pmatrix}$$

A further extension of the form (2.9) is needed to include the following model.

- SIR model with vertical transmission

A model has been proposed in [40] which extends the SIR model with vital dynamics to include vertical transmission and possible vaccination. It is assumed that b and b' are the rates of birth of uninfected and infected individuals respectively; r and r' are the corresponding death rates; v is the rate of recovery from infection; γ is the rate at which immune individuals loose immunity; q is the rate of vertical transmission ($p+q=1$); and m is the fraction of those born to uninfected parents which are immune because of vaccination, the rest going into a susceptible class. It has been assumed that the vaccine is not effective for the children of infected parents.

The ODE system which describes mathematically such a model is then the following,

$$(2.13)\qquad \begin{cases} \dfrac{dS}{dt} = -kSI + (1-m)b(S+R) + pb'I - rS + \gamma R \\ \dfrac{dI}{dt} = kSI + qb'I - r'I - vI \\ \dfrac{dR}{dt} = vI - (r+\gamma)R + mb(S+R) \end{cases}$$

In order to keep a constant total population $S+I+R=1$, it is assumed that $b=r$, $b'=r'$. In this last case the above model reduces to

$$(2.14)\qquad \begin{cases} \dfrac{dS}{dt} = -kSI + (1-m)b(1-I) + pb'I - rS + \gamma R \\ \dfrac{dI}{dt} = kSI - (pb'+v)I \end{cases}$$

If we set

$$A = \begin{pmatrix} 0 & -k \\ k & 0 \end{pmatrix}; \qquad e = \begin{pmatrix} -b-\gamma \\ -pb'-v \end{pmatrix}$$

$$c = \begin{pmatrix} (1-m)b+\gamma \\ 0 \end{pmatrix}; \qquad B = \begin{pmatrix} 0 & (m-1)b+pb'+\gamma \\ 0 & 0 \end{pmatrix}$$

system (2.14) can be written in the form

$$(2.15)\qquad \frac{dz}{dt} = diag(z)(e+Az) + c + Bz$$

which extends equation (2.9) to include the term Bz.

This kind of approach of a unifying mathematical structure of epidemic systems can be further carried out by analyzing epidemic models in two or more interacting populations.

2.2. Epidemic models with two or more interacting populations

Typical examples of epidemics which are spread by means of the interaction between different population groups are those related to venereal diseases.

Let us refer as an example to gonorrhea (due to the bacterium "Neisseria gonorrhoeae", the gonococcus).

This disease is transmitted by sexual contacts of males and females. Thus we need to consider the two interacting populations of males (1) and females (2) each of which will be divided in the two groups of susceptibles ($S_i, i = 1, 2$) and infectives ($I_i, i = 1, 2$).

We have to take into account the fact that in this case acquired immunity to reinfection is virtually non existent and hence recovered individuals pass directly back to the corresponding susceptible pool.

Death and isolation can be ignored [118].

Models of this kind are called SIS models.

2.2.1. Gonorrhea model [71, 118]

We consider here the simple gonorrhea model proposed by Cooke and Yorke [71]. It can be seen as an SIS model for two interacting populations; if we denote by $S_i, I_i, i = 1, 2$ the susceptible and the infective populations for the two groups (males and females), we have:

$$(2.16) \qquad \begin{cases} \dfrac{dS_1}{dt} = -k_{12}S_1I_2 + \alpha_1 I_1 \\ \dfrac{dI_1}{dt} = k_{12}S_1I_2 - \alpha_1 I_1 \\ \dfrac{dS_2}{dt} = -k_{21}S_2I_1 + \alpha_2 I_2 \\ \dfrac{dI_2}{dt} = k_{21}S_2I_1 - \alpha_2 I_2 \end{cases}$$

Since clearly $S_i + I_i = c_i$ (const), $i = 1, 2$, we may limit the analysis to the following system (we assume, $k_{12} = k_{21} = 1$, for simplicity)

$$(2.17) \qquad \begin{cases} \dfrac{dI_1}{dt} = -I_1I_2 - \alpha_1 I_1 + c_1 I_2 \\ \dfrac{dI_2}{dt} = -I_1I_2 - \alpha_2 I_2 + c_2 I_1 \end{cases}$$

which now can be written in the form

$$(2.18) \qquad \frac{dz}{dt} = diag(z)(e + Az) + Bz, \qquad t > 0$$

if we set $z = (I_1, I_2)^T$, and

$$A = \begin{pmatrix} 0 & -1 \\ -1 & 0 \end{pmatrix}, \quad e = \begin{pmatrix} -\alpha_1 \\ -\alpha_2 \end{pmatrix}, \quad B = \begin{pmatrix} 0 & c_1 \\ c_2 & 0 \end{pmatrix}$$

2.2.2. SIS model in two communities with migration [110]

In a SIS system with vital dynamics the population is divided into two communities; individuals migrate between the two groups. We describe each community by (S_i, I_i), $i = 1, 2$ such that

$$S_i + I_i = 1 \ , \qquad i = 1, 2 \ . \tag{2.19}$$

Hence we may limit the analysis to the following ODE system:

$$\begin{cases} \dfrac{dI_1}{dt} = k_1 I_1 (1 - I_1) - \gamma_1 I_1 - \delta_1 I_1 + \theta_1 (I_2 - I_1) \\ \dfrac{dI_2}{dt} = k_2 I_2 (1 - I_2) - \gamma_2 I_2 - \delta_2 I_2 + \theta_2 (I_1 - I_2) \end{cases} \tag{2.20}$$

Note that the migration terms $\theta_i (I_j - I_i)$, $i, j = 1, 2$, $i \neq j$, are intended to have an homogeneization effect between the two groups.

Models of this kind are used in ecological systems to describe populations that are divided in patches among which discrete diffusion occurs [148, 177, 206].

System (2.20) can be written as

$$\begin{cases} \dfrac{dI_1}{dt} = (k_1 - \gamma_1 - \delta_1 - \theta_1) I_1 - k_1 {I_1}^2 + \theta_1 I_2 \\ \dfrac{dI_2}{dt} = (k_2 - \gamma_2 - \delta_2 - \theta_2) I_2 - k_2 {I_2}^2 + \theta_2 I_1 \end{cases} \tag{2.21}$$

which can be put in the form (2.18) if we set

$$z = (I_1, I_2)^T \ ,$$

and

$$A = \begin{pmatrix} -k_1 & 0 \\ 0 & -k_2 \end{pmatrix}, \quad e = \begin{pmatrix} k_1 - \gamma_1 - \delta_1 - \theta_1 \\ k_2 - \gamma_2 - \delta_2 - \theta_2 \end{pmatrix}, \quad B = \begin{pmatrix} 0 & \theta_1 \\ \theta_2 & 0 \end{pmatrix}$$

2.2.3. SIS model for two dissimilar groups [110, 142, 218]

In this case the population is divided into two dissimilar groups because of age, social structure, space structure, etc.. The two groups may interact with each other via the infection process; e.g. the force of infection acting on the susceptibles S_1 of the first group will given by

$$g_1(I_1, I_2) = k_{11}I_1 + k_{12}I_2$$

and the analogous for the other group.

Thus the epidemic system is described by the following set of ODE's:

$$\begin{cases} \frac{dI_1}{dt} = (k_{11}I_1 + k_{12}I_2)(1 - I_1) - \gamma_1 I_1 - \delta_1 I_1 \\ \frac{dI_2}{dt} = (k_{21}I_1 + k_{22}I_2)(1 - I_2) - \gamma_2 I_2 - \delta_2 I_2 \end{cases} \tag{2.22}$$

which can be also written as

$$\begin{cases} \frac{dI_1}{dt} = (k_{11} - \gamma_1 - \delta_1) I_1 - k_{11}{I_1}^2 - k_{12}I_1I_2 + k_{12}I_2 \\ \frac{dI_2}{dt} = (k_{22} - \gamma_2 - \delta_2) I_2 - k_{22}{I_2}^2 - k_{21}I_2I_1 + k_{21}I_1 \end{cases} \tag{2.23}$$

complemented by

$$I_1 + S_1 = 1, \quad I_2 + S_2 = 1$$

System (2.23) can be put again in the form (2.18) if we define

$$A = \begin{pmatrix} -k_{11} & -k_{12} \\ -k_{21} & -k_{22} \end{pmatrix}; \; e = \begin{pmatrix} k_{11} - \gamma_1 - \delta_1 \\ k_{22} - \gamma_2 - \delta_2 \end{pmatrix}; \; B = \begin{pmatrix} 0 & k_{12} \\ k_{21} & 0 \end{pmatrix}.$$

This case is a particular case (two groups) of the more general case (n groups, $n \geq 2$) analyzed by Lajmanovich and Yorke in [142]. We shall deal with this multigroup case in Section 2.3.4 , or better in Section 4.6.1 .

2.2.4. Host - vector - host model [110]

In an SIS epidemic system with vital dynamics let us suppose that a unique vector is responsible for the spread of the disease among two different hosts.

In such a case we have three classes of infectives (two hosts and one vector). The force of infection acting on the vector susceptible population (S_2) is due to the infectives I_1 and I_3 of the host.

$$g_2(I_1, I_3) = k_{21}I_1 + k_{23}I_3$$

while the force of infection acting on the two hosts S_1 and S_3 due to the vector is given, respectively, by

$$g_1(I_2) = k_{12}I_2$$

$$g_3(I_2) = k_{32}I_2$$

As a consequence , by assuming, as usual in a SIS model, that

$$S_i + I_i = \text{const} \quad (=1), \qquad i = 1, 2, 3 \tag{2.24}$$

we have

$$\begin{cases} \dfrac{dI_1}{dt} = k_{12}I_2(1 - I_1) - \gamma_1 I_1 - \delta_1 I_1 \\ \dfrac{dI_2}{dt} = (k_{21}I_1 + k_{23}I_3)(1 - I_2) - \gamma_2 I_2 - \delta_2 I_2 \\ \dfrac{dI_3}{dt} = k_{32}I_2(1 - I_3) - \gamma_3 I_3 - \delta_3 I_3 \end{cases} \tag{2.25}$$

complemented by (2.24).

It is more convenient to rewrite system (2.24), (2.25) by emphasizing the susceptible populations $S_i = 1 - I_i$, which gives

$$\begin{cases} \dfrac{dS_1}{dt} = (-k_{12} - (\gamma_1 + \delta_1))S_1 + k_{12}S_1S_2 + (\gamma_1 + \delta_1) \\ \dfrac{dS_2}{dt} = (-k_{21} - k_{23} - (\gamma_2 + \delta_2))S_2 + k_{21}S_2S_1 + k_{23}S_2S_3 \\ \qquad\quad + (\gamma_2 + \delta_2) \\ \dfrac{dS_3}{dt} = (-k_{32} - (\gamma_3 + \delta_3))S_3 + k_{32}S_3S_2 + (\gamma_3 + \delta_3) . \end{cases} \tag{2.26}$$

System (2.26) can be put in the form (2.9) if we set

$$A = \begin{pmatrix} 0 & k_{12} & 0 \\ k_{21} & 0 & k_{23} \\ 0 & k_{32} & 0 \end{pmatrix};$$

$$e = \begin{pmatrix} -k_{12} - (\gamma_1 + \delta_1) \\ -k_{21} - k_{23} - (\gamma_2 + \delta_2) \\ -k_{32} - (\gamma_3 + \delta_3) \end{pmatrix}; \quad c = \begin{pmatrix} \gamma_1 + \delta_1 \\ \gamma_2 + \delta_2 \\ \gamma_3 + \delta_3 \end{pmatrix}.$$

2.3. The general structure

To include the models listed in Sections 2.1 and 2.2 we need to generalize (2.9) and write it in the more general form

(2.27) $$\frac{dz}{dt} = diag(z)(e + Az) + b(z)$$

where now

(2.28) $$b(z) = c + Bz$$

with

(*i*) $c \in \mathbb{R}^n_+$ a constant vector

and

(*ii*) $B = (b_{ij})_{i,j=1,\ldots,n}$ a real constant matrix such that

$$b_{ij} \geq 0, \qquad i,j = 1,\ldots,n$$

$$b_{ii} = 0, \qquad i = 1,\ldots,n$$

For system (2.27) we shall give a detailed analysis of the asymptotic behavior based on recent results due to Beretta and Capasso [28].

2.3.1. Constant total population

We consider at first the case in which the total population N is constant.

A direct consequence is that any trajectory $\{z(t), t \in \mathbb{R}_+\}$ of system (2.27) is contained in a bounded domain $\Omega \subset \mathbb{R}^n$:

(A1) Ω is positively invariant.

Because of the structure of $F : \mathbb{R}^n \to \mathbb{R}^n$ defined by

$$F(z) := diag(z)(e + Az) + b(z) \tag{2.29}$$

it is clear that $F \in C^1(\Omega)$.

We shall denote by D_i the hyperplane of $\mathbb{R}^n$:

$$D_i = \{z \in \mathbb{R}^n \mid z_i = 0\}, \qquad i = 1, \ldots, n \ .$$

Clearly, for any $i = 1, \ldots, n$, $D_i \cap \Omega$ will be positively invariant if $b_i |_{D_i} = 0$, while $D_i \cap \Omega$ will be a repulsive set whenever $b_i |_{D_i} > 0$, in which case $F(z)$ will be pointing inside Ω on D_i.

Because of the invariance of Ω and the fact that $F \in C^1(\Omega)$, standard fixed point theorems [180] (Appendix B, Section B.1) assure the existence of at least one equilibrium solution of (2.27), within Ω.

Suppose now that a strictly positive equilibrium z^* exists for system (2.27) ($z_i^* > 0, \quad i = 1, \ldots, n$):

$$diag\,(z^*)\,(e + Az^*) + b\,(z^*) = 0$$

from which we get

$$e = -Az^* - diag\left(z^{*-1}\right) b\,(z^*) \tag{2.30}$$

where we have denoted by

$$z^{*-1} := \left(\frac{1}{z_1^*}, \ldots, \frac{1}{z_n^*}\right)^T$$

By substitution into (2.27), we get

$$\begin{aligned} \frac{dz}{dt} &= diag(z)\left[A + diag\left(z^{*-1}\right) B\right](z - z^*) \\ &\quad - diag\,(z - z^*)\, diag\left(z^{*-1}\right) b(z) \end{aligned} \tag{2.31}$$

Since (2.27) is a Volterra like system we may make use of the classical Volterra-Goh Lyapunov function [96].

$$V(z) := \sum_{i=1}^{n} w_i \left(z_i - z_i^* - z_i^* \ln \frac{z_i}{z_i^*} \right), \quad z \in \mathbb{R}^{n*} \tag{2.32}$$

where $w_i > 0, \quad i = 1, \ldots, n$, are real constants (the weights).

Here we denote by

$$\mathbb{R}_+^{n*} := \{ z \in \mathbb{R}^n \mid z_i > 0, \quad i = 1, \ldots, n \},$$

and clearly

$$V : \mathbb{R}_+^{n*} \to \mathbb{R}_+ \quad .$$

The derivative of V along the trajectories of (2.27) is given by

$$\dot{V}(z) = (z - z^*)^T \; W\tilde{A}\,(z - z^*) - \sum_{i=1}^{n} w_i \frac{b_i(z)}{z_i z_i^*}(z_i - z_i^*)^2, \quad z \in \mathbb{R}_+^{n*} \tag{2.33}$$

which can be rewritten as

$$\dot{V}(z) = (z - z^*)^T \; W \left[\tilde{A} + diag \left(\frac{-b_1(z)}{z_1 z_1^*}, \ldots, \frac{-b_n(z)}{z_n z_n^*} \right) \right] (z - z^*) \tag{2.34}$$

We have denoted by $W := diag(w_1, \ldots, w_n)$, and by

$$\tilde{A} := A + diag\left(z^{*-1} \right) B \quad . \tag{2.35}$$

The structure of (2.33) and (2.34) stimulates the analysis of the following two cases:

(A) $\quad \tilde{A}$ is W-skew symmetrizable

(B) $\quad -\left[\tilde{A} + diag \left(\frac{-b_1(z)}{z_1 z_1^*}, \ldots, \frac{-b_n(z)}{z_n z_n^*} \right) \right] \in S_W \quad .$

We say that a real $n \times n$ matrix A is "skew-symmetric" if $A^T = -A$.

We say that a real $n \times n$ matrix A is W-skew symmetrizable if there exists a positive diagonal real matrix W such that WA is skew-symmetric.

We say that a real $n \times n$ matrix A is in S_W (resp. "Volterra-Lyapunov stable") if there exists a positive diagonal real matrix W such that $WA + A^T W$ is positive definite (resp. negative definite).

In case (B)

$$\dot{V}(z) \leq 0, \qquad z \in \mathbb{R}_+^{n*}$$

and the equality applies if and only if $z = z^*$. The global asymptotic stability of z^* follows from the classical Lyapunov theorem (Appendix A, Section A.5). Thus we have proved the following

Theorem 2.1. *If system (2.27) admits a strictly positive equilibrium* $z^* \in \Omega\,(z_i > 0, i = 1, \ldots, n)$ *and condition (B) applies, then* z^* *is globally asymptotically stable within* Ω. *The uniqueness of such an equilibrium point follows from the GAS.*

Consider case (A) now. Since $W\tilde{A}$ is skew-symmetric, from (2.33) we get

$$\dot{V}(z) = -\sum_{i=1}^{n} \frac{w_i b_i(z)}{z_i z_i^*}(z_i - z_i^*)^2 \tag{2.36}$$

Since $b_i(z) \geq 0$ for any $z \in \mathbb{R}_+^{n*}$, $i = 1, \ldots, n$, we have

$$\dot{V}(z) \leq 0 \quad .$$

Denote by $R \subset \Omega$ the set of points where $\dot{V}(z) = 0$; clearly

$$R = \{z \in \Omega \mid z_i = z_i^* \quad \text{if} \quad b_i(z) > 0, \quad i = 1, \ldots, n\} \tag{2.37}$$

We shall further denote by M the largest invariant subset of R. By the LaSalle Invariance Principle [145] (Appendix A, Section A.5) we may then state that every solution tends to M for t tending to infinity.

In order to give more information about the structure of M, we refer to graph theoretical arguments [205].

Since in case (A) the elements of $\tilde{A}$ have a skew-symmetric sign distribution, we can then associate a graph with $\tilde{A}$ by the following rules.

(α) each compartment $i \in \{1, \ldots, n\}$ is represented by a labelled knot denoted by

(a.1) "$\circ$" if $b_i(z) = 0 \quad \forall z \in \Omega$

(a.2) "$\bullet$" otherwise

(β) if a pair of knots (i, j) is such that $\tilde{a}_{i,j}\tilde{a}_{j,i} < 0$ then the two knots i and j are connected by an arc (see for examples Sect. 2.3.1.1).

The following lemma holds [205].

Lemma 2.2. *Assume that $\tilde{A}$ is skew-symmetrizable. If the associated graph is either*

(a) a tree and $\rho - 1$ of the terminal knots are $\bullet$

or

(b) a chain and two consecutive internal knots are $\bullet$

or

(c) a cycle and two consecutive knots are $\bullet$

then $M = \{z^*\}$ within R.

As a consequence of this lemma and the above arguments we may state the following

Theorem 2.3. *If system (2.27) admits a strictly positive equilibrium $z^* \in \Omega$ $(z_i^* > 0, \quad i = 1, \ldots, n)$ and condition (A) applies under one of the assumptions of Lemma 2.2, then the positive equilibrium z^* is GAS within Ω (again the uniqueness of z^* follows from its GAS).*

The interest of Theorems 2.1. and 2.3. lies in the fact that they provide sufficient conditions in order that an equilibrium solution of system (2.27) be globally asymptotically stable whenever we are able to show that it exists.

This will reduce a problem of GAS to an "algebraic" problem. On the other hand necessary and sufficient conditions for the existence of an equilibrium solution usually include "threshold" conditions on the parameters for the existence of such a nontrivial endemic state.

Sufficient conditions for the existence of a nontrivial endemic state are given in the following corollary of Theorems 2.1. and 2.3.

Corollary 2.4. *If the vector c in (2.28) (i) is strictly positive, then the system (2.27) admits a strictly positive equilibrium* $z^* \in \Omega_+$. *In either cases (A) and (B), the positive equilibrium* z^* *is GAS (and therefore unique) with respect to* Ω_+.

An extension of these results to the space heterogeneous case can be found in Sect. 5.6.

2.3.1.1. Case A: epidemic systems for which the matrix $\tilde{A}$ is W-skew symmetrizable

2.3.1.1.1. SIR model with vital dynamics

It is clearly seen from (2.35) that, since in this case $B = 0$, we have $\tilde{A} = A$ and $b(z) = c = \begin{pmatrix} \delta \\ 0 \end{pmatrix}$.

$\tilde{A}$ is thus skew-symmetric and the associated graph is •—o. Theorem 2.3. applies.

In this case the nontrivial equilibrium point, i.e. the nontrivial endemic state, is given by

$$S^* = \frac{\gamma + \delta}{k}; \qquad I^* = \frac{\delta}{k}\left(\frac{1}{S^*} - 1\right) \tag{2.38}$$

which exists iff

$$\sigma = \frac{k}{\gamma + \delta} > 1. \tag{2.39}$$

Note that if $\sigma \leq 1$ then the only equilibrium point of the system is $(1, 0)^T$, and this is GAS.

2.3.1.1.2. SIRS model with temporary immunity

Again in this case

$$\tilde{A} = A \quad \text{and} \quad b(z) = c$$

so that $\tilde{A}$ is skew-symmetric. The associated graph is also •–o, and Theorem 2.3. applies.

In this case the nontrivial endemic state is given by $z^* = (S^*, I^*)^T$, where

$$S^* = \frac{\gamma + \delta}{k} =: \frac{1}{\sigma}$$

$$I^* = \frac{(\delta + \alpha)(\sigma - 1)}{k + \alpha\sigma}$$

which exists iff $\sigma > 1$.

Otherwise, for $\sigma \leq 1$, the only equilibrium point of the system is $(1, 0)^T$.

2.3.1.1.3. SIR model with carriers

In this case

$$\tilde{A} = A \quad \text{and} \quad b(z) = c.$$

Since c is positive definite and $\tilde{A}$ is skew-symmetric, we may apply Corollary 2.4 to state that a unique positive equilibrium z^* exists,which is GAS with respect to the interior of

$$\Omega := \left\{ z = \left(S, \tilde{I}\right)^T \in \mathbb{R}_+^2 \mid S + \tilde{I} \leq 1 + C \right\}$$

In this case then an endemic state always exists. Its coordinates are given by [110]

$$S^* = 1 - \frac{kI^*}{\delta\sigma}$$

$$I^* = \frac{\delta}{2k}\left(\left(\sigma - 1 - C\frac{k}{\delta}\right) + \left(\left(\sigma - 1 - C\frac{k}{\delta}\right)^2 + 4C\frac{k\sigma}{\delta}\right)^{\frac{1}{2}}\right)$$

where, as usual, $\sigma := \dfrac{k}{\gamma + \delta}$.

2.3.1.1.4. SIR model with vertical transmission

In this case $b(z) = c + Bz$.

Moreover this system admits the following equilibrium point

$$(2.40)\qquad \begin{aligned} S^* &= \frac{pb' + v}{k} \\ I^* &= \frac{((1-m)b+\gamma)\,k - (b+\gamma)\,(pb'+v)}{(v+(1-m)\,b+\gamma)\,k} \end{aligned}$$

This is a nontrivial endemic state ($I^* > 0$) iff

$$(2.41)\qquad m < \frac{(b+\gamma)\,(k-pb'-v)}{bk}.$$

As a consequence

$$\tilde{A} := A + diag\left(z^{*-1}\right) B = \begin{pmatrix} 0 & k\dfrac{(m-1)b-\gamma-v}{pb'+v} \\ k & 0 \end{pmatrix}$$

Now, $(m-1)b, \quad -\gamma, \quad -v$ are all nonpositive quantities. We assume, to exclude extreme cases, that they are all negative. Thus a suitable positive diagonal matrix $W = diag(w_1 w_2)$ can be easily shown to exist, such that $W\tilde{A}$ reduces to $\begin{pmatrix} 0 & -k \\ k & 0 \end{pmatrix}$. We fall into case (A) Section 2.3.1. Since the associated graph is •—○, the endemic state (2.40) (under (2.41)) is GAS.

2.3.1.2. Case B: epidemic systems for which

$$-\left[\tilde{A} + diag\left(-\frac{b_1(z)}{z_1 z_1^*}, \ldots, \frac{b_n(z)}{z_n z_n^*}\right)\right] \in S_W.$$

2.3.1.2.1. Gonorrhea model

In this case (Eqn. (2.17)),

$$b(z) = Bz = \begin{pmatrix} c_1 I_2 \\ c_2 I_1 \end{pmatrix}, \qquad \tilde{A} = \begin{pmatrix} 0 & \dfrac{c_1 - I_1^*}{I_1^*} \\ \dfrac{c_2 - I_2^*}{I_2^*} & 0 \end{pmatrix}$$

consider the matrix

$$W\left[\tilde{A} + diag\begin{pmatrix} \dfrac{-c_1 I_2}{I_1^* I_1 - \dfrac{c_2 I_1}{I_2^* I_2}} \end{pmatrix}\right] = \begin{pmatrix} -w_1 \dfrac{c_1 I_2}{I_1^* I_1} & w_1 \dfrac{S_1^*}{I_1^*} \\ w_2 \dfrac{S_2^*}{I_2^*} & -w_2 \dfrac{c_2 I_1}{I_2^* I_2} \end{pmatrix} \tag{2.42}$$

which is a symmetric matrix if we choose $w_1 > 0$, and $w_2 > 0$ such that

$$w_2 \left(\frac{S_2^*}{I_2^*}\right) = w_1 \left(\frac{S_1^*}{I_1^*}\right)$$

The symmetric matrix (2.42) is negative definite. In fact the diagonal elements are negative and

$$\left(\frac{c_1 I_2}{I_1^* I_1} \frac{c_2 I_1}{I_1^* I_2} - \frac{S_1^* S_2^*}{I_1^* I_2^*}\right) w_1 w_2 = \frac{w_1 w_2}{I_1^* I_2^*} \left(c_1 c_2 - S_1^* S_2^*\right) > 0 \ ,$$

where the fact that $0 < S_i^* < c_i, \quad i = 1, 2$, is taken into account since $z^* = (I_1^*, I_2^*)^T$ is a positive equilibrium. Theorem 2.3 applies.

2.3.1.2.2. SIS model in two communities with migration

This model has been reduced to system (2.21). Hence

$$b(z) = Bz = \begin{pmatrix} \theta_1 I_2 \\ \theta_2 I_1 \end{pmatrix}, \qquad \text{and} \qquad \tilde{A} = \begin{pmatrix} -k_1 & \dfrac{\theta_1}{I_1^*} \\ \dfrac{\theta_2}{I_2^*} & -k_2 \end{pmatrix}$$

Let $\Omega \subset \mathbb{R}^2$ be defined as

$$\Omega := \left\{ z = (I_1, I_2)^T \in \mathbb{R}^2 \mid 0 \le I_i \le 1, \quad i = 1,2 \right\}$$

Because of Theorem 2.3, the sufficient condition for the asymptotic stability of a positive equilibrium z^*, with respect to Ω is

$$-\left[\tilde{A} + diag\left(-\frac{\theta_1 I_2}{I_1^* I_1}, -\frac{\theta_2 I_1}{I_2^* I_2} \right) \right] \in S_W$$

We can observe that

$$W\tilde{A} + diag\left(-w_1 \frac{\theta_1 I_2}{I_1^* I_1}, -w_2 \frac{\theta_2 I_1}{I_2^* I_2} \right) \tag{2.43}$$

$$= \begin{pmatrix} -w_1 \dfrac{\theta_1 I_2}{I_1^* I_1} & w_1 \dfrac{\theta_1}{I_1^*} \\ w_2 \dfrac{\theta_2}{I_2^*} & -w_2 \dfrac{\theta_2 I_1}{I_2^* I_2} \end{pmatrix} + diag\left(-k_1 w_1, -k_2 w_2\right)$$

The first matrix on the right hand side of (2.43) is symmetric if we choose $w_1 > 0, \quad w_2 = \dfrac{\theta_1 I_2^*}{\theta_2 I_1^*} w_1$.

This matrix is negative semidefinite since

$$\begin{pmatrix} \dfrac{\theta_1 I_2}{I_1^* I_1} & \dfrac{\theta_2 I_1}{I_2^* I_2} - \dfrac{\theta_1}{I_1^*} & \dfrac{\theta_2}{I_2^*} \end{pmatrix} w_1 w_2 = 0 \ .$$

Because of the presence of the diagonal negative matrix on the right hand side of (2.43), the sufficient condition of Theorem 2.3. holds true provided that $k_1, k_2 > 0$.

Under these assumptions, if a positive equilibrium z^* exists, then it is GAS within Ω.

2.3.1.2.3. SIS model for two dissimilar groups

This model has been reduced to the form (2.23).
Hence

$$b(z) \equiv Bz = \begin{pmatrix} k_{12}I_2 \\ k_{21}I_1 \end{pmatrix}, \quad \tilde{A} = \begin{pmatrix} -k_{11} & \frac{k_{12}}{I_1^*}(1-I_1^*) \\ \frac{k_{21}}{I_2^*}(1-I_2^*) & -k_{22} \end{pmatrix}$$

Consider now

$$W\left[\tilde{A} + diag\left(-\frac{k_{12}I_2}{I_1^*I_1}, -\frac{k_{21}I_1}{I_2^*I_2}\right)\right] \tag{2.44}$$

$$= \begin{pmatrix} -w_1\frac{k_{12}I_2}{I_1^*I_1} & w_1\frac{k_{12}}{I_1^*}(1-I_1^*) \\ w_2\frac{k_{21}}{I_2^*}(1-I_2^*) & -w_2\frac{k_{21}I_1}{I_2^*I_2} \end{pmatrix} + diag\left(-k_{11}w_1, -k_{22}w_2\right)$$

where the first matrix on the right hand side of (2.44) is symmetric when choosing $w_1 > 0$ and w_2 such that $\left(\frac{k_{21}}{I_2^*}\right)(1-I_2^*)\,w_2 = \left(\frac{k_{12}}{I_1^*}\right)(1-I_1^*)\,w_1$. Moreover, since $0 < I_i^* < 1, \quad i = 1, 2$, this matrix is negative definite. In fact,

$$\left(\frac{k_{12}I_2}{I_1^*I_1}\,\frac{k_{21}I_1}{I_2^*I_2} - \frac{k_{12}}{I_1^*}(1-I_1^*)\frac{k_{21}}{I_2^*}(1-I_2^*)\right)w_1w_2 > 0.$$

Hence, provided that $k_{11} \geq 0, k_{22} \geq 0$,

$$-\left[\tilde{A} + diag\left(-\frac{k_{12}I_2}{I_1^*I_1}, -\frac{k_{21}I_1}{I_2^*I_2}\right)\right] \in S_W$$

and Theorem 2.3. assures the asymptotic stability of the positive equilibrium z^* with respect to $\Omega = \left\{z \in \mathbb{R}_+^2 \mid I_i \leq 1, \quad i = 1, 2\right\}$.

2.3.1.2.4. Host - vector- host model

This model has been reduced to the form (2.26).
Hence

$$b(z) \equiv c \qquad , \qquad \tilde{A} \equiv A$$

By Corollary 2.4, since c is a positive definite vector, one positive equilibrium z^* exists in $\overset{\circ}{\Omega}$, where

$$\Omega := \left\{ z \in \mathbb{R}^3_+ \mid 0 \leq S_i \leq 1, \quad i = 1, 2, 3 \right\}.$$

$\tilde{A}$ has a symmetric sign structure. Hence, by Corollary 2.4, if

$$-\left[A + diag \left(\frac{-(\gamma_1 + \delta_1)}{S_1 S_1^*}, \frac{-(\gamma_2 + \delta_2)}{S_2 S_2^*}, \frac{-(\gamma_3 + \delta_3)}{S_3 S_3^*} \right) \right] \in S_W$$

then z^* is asymptotically stable within $\overset{\circ}{\Omega}$. If we take into account that $S_i \leq 1, \quad i = 1, 2, 3$, from (2.34) we see that a sufficient condition for the asymptotic stability of z^* is

$$-[A + diag\,(-\,(\gamma_1 + \delta_1)\,, -\,(\gamma_2 + \delta_2)\,, -\,(\gamma_3 + \delta_3))] \in S_W.$$

Accordingly, let us take

$$W\,[A + diag\,(-\,(\gamma_1 + \delta_1)\,, -\,(\gamma_2 + \delta_2)\,, -\,(\gamma_3 + \delta_3))]$$

$$= \begin{pmatrix} -\,(\gamma_1 + \delta_1)\,w_1 & k_{12} w_1 & 0 \\ k_{21} w_2 & -\,(\gamma_2 + \delta_2)\,w_2 & k_{23} w_2 \\ 0 & k_{32} w_3 & -\,(\gamma_3 + \delta_3)\,w_3 \end{pmatrix}$$

This matrix is symmetric if we choose

$$w_1 > 0, \quad w_2 = \left(\frac{k_{12}}{k_{21}} \right) w_1, \quad w_3 = \left(\frac{k_{23}}{k_{32}} \right) \left(\frac{k_{12}}{k_{21}} \right) w_1.$$

It is negative definite if

$$[(\gamma_1 + \delta_1)\,(\gamma_2 + \delta_2) - k_{12} k_{21}]\, w_1 w_2 > 0,$$

(2.45)
$$-[\,(\gamma_1 + \delta_1)\,(\gamma_2 + \delta_2)\,(\gamma_3 + \delta_3) - (\gamma_3 + \delta_3)\,k_{12} k_{21}$$
$$-\,(\gamma_1 + \delta_1)\,k_{23} k_{32}] w_1 w_2 w_3 < 0$$

We can observe that, if inequalities in (2.45) hold true, then

$$[(\gamma_2 + \delta_2)\,(\gamma_3 + \delta_3) - k_{23} k_{32}]\, w_2 w_3 > 0\ .$$

Hence (2.45) is the sufficient condition for the asymptotic stability (and uniqueness) of the positive equilibrium z^* within $\overset{\circ}{\Omega}$.

From (2.26) the positive equilibrium z^* has the following components:

$$(2.46)\qquad S_1^* = \frac{\gamma_1+\delta_1}{k_{12}(1-S_2^*)+(\gamma_1+\delta_1)} \;, \qquad S_3^* = \frac{\gamma_3+\delta_3}{k_{32}(1-S_2^*)+(\gamma_3+\delta_3)} \;,$$

and S_2^* is a solution of

$$(2.47)\qquad (1-S_2)\left\{p(1-S_2)^2+q(1-S_2)+r\right\}=0 \;,$$

where

$$p = k_{12}k_{32}\left[(k_{21}+k_{23})+(\gamma_2+\delta_2)\right] \;,$$

$$q = k_{32}\left[(\gamma_1+\delta_1)(\gamma_2+\delta_2)-k_{12}k_{21}\right]+k_{12}\left[(\gamma_2+\delta_2)(\gamma_3+\delta_3)-k_{23}k_{32}\right] + k_{12}k_{21}(\gamma_3+\delta_3)+k_{23}k_{32}(\gamma_1+\delta_1)$$

$$r = (\gamma_1+\delta_1)(\gamma_2+\delta_2)(\gamma_3+\delta_3)-(\gamma_3+\delta_3)k_{12}k_{21}-(\gamma_1+\delta_1)k_{23}k_{32} \;.$$

It is to be noticed that when (2.45) holds true, then $q>0$, $r>0$, thus assuring that the unique asymptotically stable equilibrium is such that $S_2^* = 1$, i.e. $z^* = (1,1,1)^T$.

When (2.45) fails to hold, by (2.47) we have another positive equilibrium for which $S_2^* < 1$ and its remaining components are given by (2.46).

To study the asymptotic stability of this equilibrium we can remember that $I_i + S_i = 1$, $i = 1,2,3$, thus assuring to have a positive equilibrium $z^* = (I_1^*, I_2^*, I_3^*)^T$, $0 < I_i^* < 1$, $i = 1,2,3$ within the subset $\overline{\Omega} = \left\{z \in \mathbb{R}_+^3 : I_i \le 0, \quad i=1,2,3\right\}$. In the old variables I_i, $i = 1,2,3$, the positive equilibrium becomes the origin and the ODE system (2.25) can be arranged in this form:

$$\begin{aligned}
\frac{dI_1}{dt} &= -(\gamma_1+\delta_1)I_1 - k_{12}I_1I_2 + k_{12}I_2 \;,\\
\frac{dI_2}{dt} &= -(\gamma_2+\delta_2)I_2 - k_{21}I_2I_1 - k_{23}I_2I_3 + (k_{21}I_1 + k_{23}I_3) \;,\\
\frac{dI_3}{dt} &= -(\gamma_3+\delta_3)I_3 - k_{32}I_3I_2 + k_{32}I_2
\end{aligned}$$

so that

$$e = \begin{pmatrix} -(\gamma_1+\delta_1)\\ -(\gamma_2+\delta_2)\\ -(\gamma_3+\delta_3)\end{pmatrix}, \qquad A = \begin{pmatrix} 0 & -k_{12} & 0\\ -k_{21} & 0 & -k_{23}\\ 0 & -k_{32} & 0\end{pmatrix}$$

$$c = 0, \qquad B = \begin{pmatrix} 0 & k_{12} & 0\\ k_{21} & 0 & k_{23}\\ 0 & k_{32} & 0\end{pmatrix}$$

Thus

$$b(z) = Bz, \qquad \tilde{A} = \begin{pmatrix} 0 & \dfrac{k_{12}S_1^*}{I_2^*} & 0 \\ \dfrac{k_{21}S_2^*}{I_2^*} & 0 & \dfrac{k_{23}S_2^*}{I_2^*} \\ 0 & \dfrac{k_{32}S_3^*}{I_3^*} & 0 \end{pmatrix}$$

For the asymptotic stability of $z^* = (I_1^*, I_2^*, I_3^*)^T$ within $\overline{\Omega}$ we can apply Theorem 2.3 by requiring that

$$-\left[\tilde{A} + diag\left(-\frac{b_1(z)}{I_1 I_1^*}, -\frac{b_2(z)}{I_2 I_2^*}, -\frac{b_3(z)}{I_3 I_3^*}\right)\right] \in S_W .$$

Hence consider

$$W\left[\tilde{A} + diag\left(-\frac{b_1(z)}{I_1 I_1^*}, -\frac{b_2(z)}{I_2 I_2^*}, -\frac{b_3(z)}{I_3 I_3^*}\right)\right] \tag{2.48}$$

$$= \begin{pmatrix} -\dfrac{k_{12}I_2}{I_1 I_1^*}w_1 & \dfrac{k_{12}S_1^*}{I_1^*}w_1 & 0 \\ \dfrac{k_{21}S_2^*}{I_2^*}w_2 & -\dfrac{(k_{21}I_1 + k_{23}I_3)}{I_2 I_2^*}w_2 & \dfrac{k_{23}S_2^*}{I_2^*}w_2 \\ 0 & \dfrac{k_{32}S_3^*}{I_3^*}w_3 & -\dfrac{k_{32}I_2}{I_3 I_3^*}w_3 \end{pmatrix}$$

this matrix is symmetric if we choose

$$w_1 > 0 \ , \quad w_2 = \left(\frac{k_{12}S_1^*}{k_{21}S_2^*}\right)\left(\frac{I_2^*}{I_1^*}\right) w_1 \ , \quad w_3 = \left(\frac{k_{23}S_2^*}{k_{32}S_3^*}\right)\left(\frac{I_3^*}{I_2^*}\right) w_2 \ .$$

To apply Theorem 2.3 we must require that the symmetric matrix (2.48) be negative definite. Since the diagonal elements are negative, the sufficient condition is

$$\left[\frac{k_{12}I_2}{I_1} \ \frac{(k_{21}I_1 + k_{23}I_3)}{I_2^*} - k_{21}S_1^* k_{21}S_2^*\right] \frac{w_1 w_2}{I_1^* I_2^*} > 0 \ ,$$

$$\left[-\frac{k_{12}I_2}{I_1} \ \frac{(k_{21}I_1 + k_{23}I_3)}{I_2} \ \frac{k_{32}I_2}{I_3} + \frac{k_{32}I_2}{I_3} k_{12}S_1^* k_{21}S_2^* \right. \tag{2.49}$$

$$\left. + \frac{k_{12}I_2}{I_1} k_{23}S_2^* k_{32}S_3^*\right] \frac{w_1 w_2 w_3}{I_1^* I_2^* I_3^*} < 0 \ .$$

Now we observe that the sufficient condition (2.49) is always met by a positive equilibrium $z^* \in \overline{\Omega}$.

In fact

$$\frac{k_{12}I_2}{I_1}\ \frac{(k_{21}I_1+k_{23}I_3)}{I_2}-k_{12}S_1^*k_{21}S_2^*>\frac{k_{12}I_2}{I_1}\ \frac{k_{21}I_1}{I_2}-k_{12}k_{21}=0$$

and

$$-\frac{k_{12}I_2}{I_1}\ \frac{k_{21}I_1}{I_2}\ \frac{k_{32}I_2}{I_3}+\frac{k_{32}I_2}{I_3}\ k_{12}S_1^*k_{21}S_2^*-\frac{k_{12}I_2}{I_1}\ \frac{k_{23}I_3}{I_2}\ \frac{k_{32}I_2}{I_3}$$
$$+\frac{k_{12}I_2}{I_1}\ k_{23}S_2^*k_{32}S_3^*=$$
$$=\frac{I_2}{I_3}k_{32}\left(-k_{12}k_{21}+k_{12}S_1^*k_{21}S_2^*\right)+\frac{k_{12}I_2}{I_1}\left(-k_{23}k_{32}+k_{23}S_2^*k_{32}S_3^*\right)<0\ ,$$

where, when proving the inequalities, we have taken into account that $S_i^*<1,\quad i=1,2,3$.

Hence we can conclude for the host-vector-host model that

Proposition 2.5. *If the sufficient condition (2.45) holds true, then the origin is asymptotically stable with respect to $\overline{\Omega}$. Otherwise besides the origin a positive equilibrium $z^*\in\overline{\Omega}$ exists which is GAS in $\overset{\circ}{\Omega}$.*

2.3.2. Nonconstant total population

In some relevant cases the total population

$$N(t)=\sum_{i=1}^{n} z_i(t) \tag{2.50}$$

of the epidemic system is not a constant, but rather a dynamical variable. We shall consider in the sequel specific examples of this kind.

A first model is the parasite-host system studied by Levin and Pimentel in [151]:

$$\begin{cases}\dfrac{dx}{dt}=(r-k)x-Cxy-Cxv+ry+rv\ ,\\[2mm] \dfrac{dy}{dt}=-(\beta+k)\,y+Cxy-CSyv\ ,\\[2mm] \dfrac{dv}{dt}=-(\beta+k+\sigma)\,v+Cxv-CSyv\end{cases} \tag{2.51}$$

The two cases $r<k$ and $r>\beta+k+\sigma$ do not give rise to nontrivial equilibrium solutions. We shall then restrict our analysis to the case $\beta+\sigma+k>r>k$ in which there is an equilibrium at

$$x^*=\frac{r}{C}\frac{\sigma}{\sigma-S(r-k)},\quad y^*=\frac{\beta+k+\sigma}{CS}-\frac{1}{S}x^*,\quad v^*=\frac{1}{S}x^*-\frac{\beta+k}{CS} \tag{2.52}$$

The equilibrium $z^* = (x^*, y^*, v^*)$ is feasible, i.e. its components are positive if

$$\frac{r}{\beta+k+\sigma} < 1 - \frac{S(r-k)}{\sigma} < \frac{r}{\beta+k}\ , \tag{2.53}$$

If $\sigma < \sigma_1$ where σ_1 is such that

$$\frac{r}{\beta+k+\sigma_1} = 1 - \frac{S(r-k)}{\sigma_1}\ , \tag{2.54}$$

the first inequality in (2.53) is violated and only a partially feasible equilibrium is present given by

$$x^* = \frac{\beta+k+\sigma}{C}, \quad y^* = 0, \quad v^* = \frac{r-k}{\beta+k+\sigma+r}x^* \tag{2.55}$$

since $r < \beta+k+\sigma$. If $\sigma = \sigma_1$ then (2.52) coalesces in (2.55). If $r < \beta+k$ and $\sigma > \sigma_2$, where σ_2 is such that

$$1 - \frac{S(r-k)}{\sigma_2} = \frac{r}{\beta+k} \tag{2.56}$$

then the second inequality in (2.53) is violated and only a partially feasible equilibrium is present, given by

$$x^* = \frac{\beta+k}{C}, \quad y^* = \frac{r-k}{\beta+k-r}x^*, \quad v^* = 0 \tag{2.57}$$

since $r > k$. If $\sigma = \sigma_2$ then (2.52) coalesces in (2.57).

Concerning system (2.51), if we denote by $z = (x, y, v)^T$ and set

$$A = \begin{pmatrix} 0 & -C & -C \\ C & 0 & -CS \\ C & CS & 0 \end{pmatrix}; \quad e = \begin{pmatrix} r-k \\ -(\beta+k) \\ -(\beta+k+\sigma) \end{pmatrix}$$

$$c = 0, \quad B = \begin{pmatrix} 0 & r & r \\ 0 & 0 & 0 \\ 0 & 0 & 0 \end{pmatrix}$$

we may reduce it again to the general structure (2.27), but in this case

$$\frac{dN}{dt} = (r-k)N(t) \tag{2.58}$$

and the evolution of system (2.51) has to be analyzed in the whole $\mathbb{R}^3_+$.

Local stability results were already given in [151]. Here we shall study global asymptotic stability of the feasible or partially feasible equilibrium by the Beretta-Capasso approach (see Section 2.3.2.1).

A second model that we shall analyze is the SIS model with vital dynamics, which is proposed by Anderson and May [8]

$$\begin{aligned}\frac{dS}{dt} &= (r-b)S - \rho SI + (\mu + r)\, I \\ \frac{dI}{dt} &= -\,(\theta + b + \mu)\, I + \rho SI\end{aligned} \tag{2.59}$$

If we denote by $z = (S, I)^T$ and set

$$A = \begin{pmatrix} 0 & -\rho \\ \rho & 0 \end{pmatrix}; \quad e = \begin{pmatrix} r-b \\ -\,(\theta+b+\mu) \end{pmatrix}$$

$$c = 0; \quad B = \begin{pmatrix} 0 & \mu + r \\ 0 & 0 \end{pmatrix}$$

we go back to system (2.27). In this case

$$\frac{dN}{dt}(t) = (r-b)N(t) - \theta I(t)$$

Other examples will be discussed later. It is clear that if the total population is a dynamical variable rather than a specified constant, we need to drop assumption (A1) in Section 2.3.1.

For these systems the accessible space is the whole nonnegative orthant $\mathbb{R}^n_+$ of the Euclidean space. We cannot apply then the standard fixed point theorems.

We can only assume that

$$(A2) \qquad \mathbb{R}^n_+ \quad \text{is positively invariant.}$$

We shall give now more extensive treatment of system (2.27) including the possibility of partially feasible equilibrium points.

We shall say that z^* is a partially feasible equilibrium whenever a nonempty proper subset of its components are zero. If we denote by $N = \{1, \ldots, n\}$, we mean that a set $I \subset N$ exists, such that $I \neq \emptyset$, $I \neq N$ and $z_i^* = 0$ for any $i \in I$.

Assume from now on that this is the case; given the matrices

$$A = (a_{ij})_{i,j=1,\ldots,n} \quad \text{and} \quad B = (b_{ij})_{i,j=1,\ldots,n}$$

in system (2.27), we define a new matrix

$$\tilde{A} = (\tilde{a}_{ij})_{i,j=1,\ldots,n}$$

as follows

$$\tilde{a}_{ij} = a_{ij} + \frac{b_{ij}}{z_i^*}, \quad i \in N - I, \quad j \in N$$

$$\tilde{a}_{ij} = a_{ij}, \qquad \text{otherwise.}$$

With the above notations in mind, system (2.27) can be rewritten as

$$\text{(2.60}a\text{)} \qquad \frac{dz_i}{dt} = z_i \sum_{j \in N} \tilde{a}_{ij} \left(z_j - z_j^*\right) - \frac{(z_i - z_i^*)}{z_i^*} b_i(z), \quad i \in N - I$$

$$\text{(2.60}b\text{)} \qquad \frac{dz_i}{dt} = z_i \left(e_i + \sum_{j \in N} a_{ij} z_j \right), \quad i \in I$$

We introduce a new Lyapunov function suggested by Goh [94, 95, 96]

$$\text{(2.61)} \qquad V(z) = \sum_{i \in N-I} w_i \left(z_i - z_i^* - z_i^* \ln \frac{z_i}{z_i^*} \right) + \sum_{i \in I} w_i z_i$$

where, as usual $w_i > 0, \quad i = 1, \ldots, n$.

Clearly $V \in C^1\left(R_I^n\right)$, where we define

$$\text{(2.62)} \qquad R_I^n := \{ z \in \mathbb{R}^n \mid z_i > 0, i \in N - I; \quad z_i \geq 0, \; i \in I \}$$

Let R be the subset of R_I^n defined as follows

$$\text{(2.63)} \quad R := \{ z \in R_I^n \mid z_i = 0, i \in I, z_i = z_i^* \text{ for any } i \in N - I \text{ s.t. } b_i(z) > 0 \}$$

and let M be the largest invariant subset of R with respect to the system (2.27).

On account of (2.60) the time derivative of V along the trajectories of system (2.27) is given by

$$\dot{V}(z) = \sum_{i \in N-I} w_i \frac{(z_i - z_i^*)}{z_i} \left\{ z_i \sum_{j \in N} \tilde{a}_{ij} \left(z_j - z_j^*\right) - \frac{(z_i - z_i^*)}{z_i^*} b_i(z) \right\}$$
$$+ \sum_{i \in I} w_i z_i \left(e_i + \sum_{j \in N} a_{ij} z_j \right)$$

or, in matrix notation ($W = diag\,(w_i, \quad i = 1, \ldots, n)$)

$$\text{(2.64)} \qquad \begin{aligned} \dot{V}(z) &= (z - z_i^*)^T \, W \tilde{A} \, (z - z^*) - \sum_{i \in N-I} w_i \frac{b_i(z)}{z_i z_i^*} \left(z_i - z_i^*\right)^2 \\ &+ \sum_{i \in I} w_i \left\{ z_i \left(e_i + \sum_{j \in N} a_{ij} z_j \right) + b_i(z) \right\} . \end{aligned}$$

It is clear that

$$R = \left\{ z \in R_I^n \mid \dot{V}(z) = 0 \right\} .$$

We are now in a position to state the following

Theorem 2.6. *Let z^* be a partially feasible equilibrium of system (2.27), with $z_i^* = 0$ for $i \in I \subset N, \quad I \neq \emptyset, I \neq N$. Assume that*

(a) $\tilde{A}$ is W-skew symmetrizable

(b) $e_i + \sum_{j \in N} a_{ij} z_j^* \leq 0, \quad i \in I$

(c) $b_i(z) \equiv 0, \quad i \in I$

(d) $M \equiv \{z^*\}$

Then z^* is globally asymptotically stable within R_I^n.

Proof. Since $\tilde{A}$ is W-skew symmetrizable, the first term in (2.64) vanishes. By the assumptions (b), $\dot{V}(z) \leq 0$ in R_I^n. We can then apply LaSalle Invariance Principle [145, Theorem VI Sect. 13 (see also Appendix A, Section A.5)],to state that z^* is GAS in R_I^n.

A natural consequence of Theorem 2.6 is the following

Corollary 2.7. *Let z^* be a feasible equilibrium of (2.27) and assume that $\tilde{A}$ is W-skew symmetrizable. If $M \equiv \{z^*\}$ then z^* is globally asymptotically stable within $\mathbb{R}_+^{n*}$.*

Corollary 2.7 can be seen as a new formulation of Theorem 2.3 in the case in which $(A1)$ is substituted by $(A2)$.

Under the same conditions of this corollary we can also observe that if the graph associated with $\tilde{A}$ by means of (α) and (β) satisfies anyone of the hypotheses in Lemma 2.2, then within R we have $M \equiv \{z^*\}$.

We can now solve the two models presented in Section 2.3.1.

2.3.2.1. The parasite-host system [151]

Consider the case in which the equilibrium (2.52) is feasible, i.e. $z^* \in \mathbb{R}^3_+$. Then

$$(2.65) \qquad b(z) \equiv Bz, \quad \tilde{A} = \begin{pmatrix} 0 & -\left(C - \frac{r}{x^*}\right) & -\left(C - \frac{r}{x^*}\right) \\ C & 0 & -CS \\ 0 & CS & 0 \end{pmatrix}$$

where z is a vector $z = (x, y, v)^T$ belonging to the non-negative orthant $\mathbb{R}^3_+$. Since $C - \dfrac{r}{x^*} = CS\dfrac{(r-k)}{\sigma}$ provided that $r > k$, matrix $\tilde{A}$ is W-skew symmetrizable by the diagonal positive matrix $W = diag\,(w_1, w_2, w_3)$, where $w_1 = \dfrac{\sigma}{S(r-k)}$, $\quad w_2 = w_3 = 1$. In fact, we obtain

$$W\tilde{A} = \begin{pmatrix} 0 & -C & -C \\ C & 0 & -CS \\ C & CS & 0 \end{pmatrix}$$

Now we are in position to apply Corollary 2.7.

Since $b(z) = (r(y+z), 0, 0)^T$, the subset of all points within $\mathbb{R}^{n*}_+$ where we have $\dot{V}(z) = 0$, is

$$R = \left\{ z \in \mathbb{R}^n_+ \mid x = x^* \right\}$$

Now we look for the largest invariant subset M within R. Since $x = x^*$ for all t, $\left.\dfrac{dx}{dt}\right|_R = 0$, and from the first of the Eqns. (2.51) we obtain

$$(y+v)\mid_R = \frac{r-k}{C - \dfrac{r}{x^*}} = \frac{\sigma}{CS}, \qquad \text{for all t.}$$

Therefore, $\left.\dfrac{d(y+v)}{dr}\right|_R = 0$, and by the last two Eqns. (2.51) we obtain

$$z\mid_R = \frac{1}{\sigma}\left\{[Cx^* - (\beta+k)]\,[(y+v)]_R\right\} = \frac{1}{CS}\,[Cx^* - (\beta+k)] = \frac{x^*}{S} - \frac{\beta+k}{CS}$$

Then, by taking into account (2.52) we have $z\mid_R \equiv z^*$.

Immediately follows

$$y\,|_R = \frac{\sigma}{CS} - v^* = \frac{\beta + k + \sigma}{CS} - \frac{x^*}{S} \ ,$$

i.e. $y\,|_R = y^*$. Then the largest invariant set M within R is z^*. From Corollary 2.7 it follows the global asymptotic stability of the feasible equilibrium (2.52) within $\mathbb{R}^{3*}_+$.

It is to be noticed that the only assumptions made in this proof are $r > k$ and that equilibrium (2.52) is feasible. Under these assumptions we exclude that unbounded solutions may exist.

Suppose that $\sigma \leq \sigma_1$, i.e. the equilibrium (2.52) is not feasible and we get the partially feasible equilibrium (2.55) which belongs to

$$R_2^3 = \left\{z \in \mathbb{R}^3 \mid z_i > 0, \quad i = 1, 3, \quad z_2 \geq 0\right\}$$

In order to apply Theorem 2.6, hypotheses (a) and (b) must be verified. Concerning hypothesis (a), we have

$$-(\beta + k) + cx^* - cSv^* \leq 0 \ , \tag{2.66}$$

from which, by taking into account (2.55), we obtain

$$1 - \frac{S(r-k)}{\sigma} \leq \frac{r}{\beta + k + \sigma} \ , \tag{2.67}$$

Inequality (2.67) is satisfied in the whole range $\sigma \leq \sigma_1$ within which the partially feasible equilibrium (2.55) occurs. When $\sigma = \sigma_1$ the equality applies in (2.53). Hypothesis (b) is satisfied because $b(z) = (r(y+v), 0, 0)^T$ and therefore $b_2(z) \equiv 0$. Concerning hypothesis (c), consider first the case $\sigma < \sigma_1$, i.e. the inequality applies in (2.53).

Then the subset (2.63) is

$$R = \left\{z \in R_2^3 \mid y = 0, \quad x = x^*\right\} .$$

Now we look for the largest invariant subset M within R.

Since $x = x^*, y = 0$ for all t, $\left.\dfrac{dx}{dt}\right|_R = 0$, and from the first of equation (2.51) we get

$$v\,|_R = \frac{r-k}{C - \dfrac{r}{x^*}} \quad \text{where} \quad x^* = \frac{\beta+k+\sigma}{C} \ .$$

Therefore, we obtain $v\,|_R = \left[\dfrac{(r-k)}{(\beta+k+\sigma-r)}\right] x^*$, i.e. $v\,|_R \equiv v^*$. Thus the largest invariant set within R is

$$z^* = \left(x^* = \frac{\beta+k+\sigma}{C}, \quad y^* = 0, \quad v^* = \frac{r-k}{\beta+k+\sigma-r} x^* \right)^T \ .$$

When $\sigma = \sigma_1$, then equality applies in (2.53), and (2.63) becomes

$$R = \left\{ z \in R_2^3 \mid x = x^* \right\} .$$

In this case, we have already proven that $M \equiv \{z^*\}$. Hence hypothesis (c) is satisfied. Then by Theorem 2.6 the partially feasible equilibrium (2.55) is globally asymptotically stable with respect to R_2^3.

If $r < \beta + k$ and $\sigma \geq \sigma_2$, then the partially feasible equilibrium (2.57) occurs. This equilibrium belongs to

$$R_3^3 = \left\{ z \in \mathbb{R}_+^3 \mid z_i > 0, \quad i = 1,2; \quad z_3 \geq 0 \right\} .$$

Hypothesis (a) of Theorem 2.6 requires

$$-(\beta+k+\sigma) + Cx^* + CSy^* \leq 0 \ , \tag{2.68}$$

from which, by taking into account (2.57), we obtain

$$1 - \frac{S(r-k)}{\sigma} \geq \frac{r}{\beta+k} \ . \tag{2.69}$$

This inequality is satisfied in the whole range of existence of the equilibrium (2.57), i.e. for all $\sigma \geq \sigma_2$.

When $\sigma = \sigma_2$, the equality applies in (2.69). Hypothesis (b) of Theorem 2.6 is obviously satisfied. Concerning hypothesis (c), at first we consider the case in which $\sigma > \sigma_2$. Therefore, the inequality applies in (2.68) and the subset (2.63) of R_3^3 is

$$R = \left\{ z \in R_3^3 \mid v = 0, x = x^* \right\}.$$

From (2.57), we are ready to prove that $M \equiv \{z^*\}$. When $\sigma = \sigma_2$, R becomes

$$R = \left\{ z \in R_3^3 \mid z = x^* \right\},$$

and we have already proven that $M \equiv \{z^*\}$. Hypothesis (c) is satisfied. Also, in this case Theorem 2.6 assures the global asymptotic stability of the partially feasible equilibrium (2.57) with respect to R_3^3.

Extensions of this model, which have raised further open mathematical problems, are due to Levin [149, 150].

2.3.2.2. An SIS model with vital dynamics

Provided that $r > b, \quad \theta > r - b$, system (2.59) has the feasible equilibrium $z^* \in \mathbb{R}_+^{2*}$:

$$S^* = \frac{\theta + b + \mu}{\rho}, \quad I^* = \frac{r - b}{\theta + b - r} S^*. \tag{2.70}$$

When $r \leq b$, or $r > \theta + b$, the equilibrium (2.70) is not feasible and the only equilibrium of (2.59) is the origin.

Here $b(z) \equiv Bz = ((\mu + r) I, 0)^T$. When z^* is a feasible equilibrium the matrix $\tilde{A} = A + diag\left(z^{*-1}\right) B$ is given by

$$\tilde{A} = \begin{pmatrix} 0 & -\left(\rho - \dfrac{\mu + r}{S^*}\right) \\ \rho & 0 \end{pmatrix}$$

Since $S^* = \dfrac{(\theta + b + \mu)}{\rho}$, provided that $\theta > r - b$ the matrix $\tilde{A}$ is skew-symmetrizable. Because $b_1(z) \geq 0$, the graph associated with $\tilde{A}$ is •—o, and by Corollary 2.7 the global asymptotic stability of z^* with respect to $\mathbb{R}_+^2$ follows.

When $r \leq b, \quad r > \theta + b$ Theorem 2.6 cannot be applied to study attractivity of the origin because hypothesis (b) is violated.

2.3.2.3. An SIRS model with vital dynamics in a population with varying size [44]

As a generalization of the model discussed in Sect. 2.1.2 and Sect. 2.3.1.1.2, in [44] Busenberg and van den Driessche propose the following SIRS model

$$(2.71)\qquad \begin{cases} \dfrac{dS}{dt} = bN - dS - \dfrac{\lambda}{N}IS + eR \\ \dfrac{dI}{dt} = -(d+\epsilon+c)I + \dfrac{\lambda}{N}IS \\ \dfrac{dR}{dt} = -(d+\delta+f)R + cI \end{cases}$$

for $t > 0$, subject to suitable initial conditions.

In this case the evolution equation for the total population N is the following one,

$$(2.72)\qquad \frac{dN}{dt} = (b-d)N - \epsilon I - \delta R\,, \qquad t > 0\ .$$

We may notice that whenever $b \neq d$, N is a dynamical variable. It is then relevant to take it into explicit account in the force of infection.

If we take into account the discussion in [110] and [118], we may realize that also model (8)-(10) in [6] should be rewritten as (2.71).

The biological meaning of the parameters in (2.71) is the following :

$b =$ per capita birth rate
$d =$ per capita disease free death rate
$\epsilon =$ excess per capita death rate of infected individuals
$\delta =$ excess per capita death rate of recovered individuals
$c =$ per capita recovery rate of infected individuals
$f =$ per capita loss of immunity rate of recovered individuals
$\lambda =$ effective per capita contact rate of infective individuals with respect to other individuals.

Clearly (2.72) implies that for $b \leq d$, $N(t)$ will tend to zero so that the only possible asymptotic state for (S, I, R) is $(0, 0, 0)$.

On the other hand, for $b > d$, N may become unbounded, and the previous methods cannot directly be applied. We shall then follow the approach proposed in [44].

As usual, we may refer to the fractions

$$s(t) = \frac{S(t)}{N(t)} \; ; \quad i(t) = \frac{I(t)}{N(t)} \; ; \quad r(t) = \frac{R(t)}{N(t)} \; , \qquad t \geq 0 \tag{2.73}$$

so that

$$s(t) + i(t) + r(t) = 1 \; , \qquad t \geq 0 \; . \tag{2.74}$$

But, being $N(t)$ a dynamical variable, going from the evolution equations (2.71) for S, I, R to the evolution equations for s, i, r we need to take (2.72) into account; we have then

$$\begin{cases} \dfrac{ds}{dt} = b - bs + fr - (\lambda - \epsilon)si + \delta sr \\ \dfrac{di}{dt} = -(b + c + \epsilon)i + \lambda si + \epsilon i^2 + \delta ir \\ \dfrac{dr}{dt} = -(b + f + \delta)r + ci + \epsilon ir + \delta r^2 \end{cases} \tag{2.75}$$

for $t > 0$.

The feasibility region of system (2.75) is now

$$\mathcal{D} := \{(s, i, r)^T \in \mathbb{R}^3_+ \mid s + i + r = 1\} \; , \tag{2.76}$$

and it is not difficult to show that it is an invariant region for (2.75).

The trivial equilibrium $(1, 0, 0)^T$ (disease free equilibrium) always exists; we shall define

$$\mathcal{D}_o := \mathcal{D} - \{(1, 0, 0)^T\} \; . \tag{2.77}$$

Our interest is to give conditions for the existence and stability of nontrivial endemic states $z^* := (s^*, i^*, r^*)^T$ such that $i^* > 0$.

This is the content of the main theorem proven in [44]. The authors make use of the following "threshold parameters"

(2.78)
$$R_o := \frac{\lambda}{b+c+\epsilon}$$

(2.79)
$$R_1 := \begin{cases} \dfrac{b}{d}, & \text{if} \quad R_o \leq 1 \\[2ex] \dfrac{b}{d+\epsilon i^* + \delta r^*}, & \text{if} \quad R_o > 1 \end{cases}$$

(2.80)
$$R_2 := \begin{cases} \dfrac{\lambda}{c+d+\epsilon}, & \text{if} \quad R_o \leq 1 \\ \dfrac{\lambda s^*}{c+d+\epsilon}, & \text{if} \quad R_o > 1 \; . \end{cases}$$

Theorem 2.8. [44] *Let $b, c > 0$, and all other parameters be non negative.*

a) If $R_o \leq 1$ then $(1,0,0)^T$ is GAS in $\mathcal{D}$
If $R_o > 1$ then $(1,0,0)^T$ is unstable

b) If $R_o > 1$ then a unique nontrivial endemic state exists $(s^*, i^*, r^*)^T$ in $\mathring{\mathcal{D}}$ which is GAS in $\mathring{\mathcal{D}}$.

Proof. It is an obvious consequence of (2.75) that the trivial solution $z^o := (1,0,0)^T$ always exists. The local stability of z^o for system (2.75) is governed by the Jacobi matrix (let $\gamma = b + c + \epsilon$)

(2.81)
$$J(z^o) = \begin{pmatrix} -b & -\lambda+\epsilon & f+\delta \\ 0 & \lambda-\gamma & 0 \\ 0 & c & -(b+f+\delta) \end{pmatrix}$$

whose eigenvalues are

(2.82)
$$(\lambda_1, \lambda_2, \lambda_3) = (-b, \gamma(R_o - 1), -(b+f+\delta)) \; .$$

Hence, if $R_o < 1$ all eigenvalues are negative and z^o is LAS. On the other hand if $R_o > 1$, $\lambda_2 > 0$ and z^o is unstable.

It can be easily seen that if $R_o \leq 1$ no nontrivial endemic state $z^* \in \mathcal{D}_o$ may exist.

By using the relation $s = 1 - i - r$ we may refer to the reduced system

$$\text{(2.83)} \qquad \begin{cases} \dfrac{di}{dt} = \gamma(R_o - 1)i - (R_o\gamma - \epsilon)i^2 - (R_o\gamma - \delta)ir \\ \dfrac{dr}{dt} = -(b + f + \delta)r + ci + \epsilon ir + \delta r^2 \end{cases}$$

whose admissible region is

$$\text{(2.84)} \qquad \mathcal{D}_1 := \{(i, r)^T \in \mathbb{R}^2_+ \mid i + r \leq 1\}$$

For the planar system (2.83) $\mathcal{D}_1$ is a bounded invariant region which cannot contain any other equilibrium point than $(0,0)^T$.

On the other hand $(0,0)^T$ is LAS in $\mathcal{D}$. Suppose it is not GAS, then for an initial condition outside a suitably chosen neighborhood of $(0,0)^T$, the corresponding orbit should remain in a bounded region which does not contain equilibrium points. By the Poincaré-Bendixon theorem, this orbit should spiral into a periodic solution of system (2.83). But in [44] it is proven that system (2.75) has no periodic solutions, nor homoclinic loops in $\mathcal{D}$, so this will be the case for system (2.83) in $\mathcal{D}_1$, and this leads to a contradiction.

The same holds for $R_o = 1$, so that part a) of the theorem is completely proven.

As far as part b) is concerned, from system (2.83) we obtain that a nontrivial equilibrium solution ($i^* > 0$) must satisfy

$$\text{(2.85)} \qquad \begin{cases} \gamma(1 - R_o) + (\lambda - \epsilon)i + (\lambda - \delta)r = 0 \\ -(b + f + \delta)r + ci + \epsilon ir + \delta r^2 = 0 \end{cases}$$

which is proven to have a unique nontrivial solution $(i^*, r^*)^T \in \mathring{\mathcal{D}}_1$.

The local stability of this equilibrium is governed by the matrix

$$J(i^*, r^*) = \begin{pmatrix} -(R_o\gamma - \epsilon)i^* & -(R_o\gamma - \delta)i^* \\ c + \epsilon r^* & -(b + f + \delta)\epsilon i^* + 2\delta r^* \end{pmatrix}$$

By the Routh-Hurwitz criterion it is not difficult to show that $(i^*, r^*)^T$ is LAS.

Again the Poincaré-Bendixon theory, together with the nonexistence of periodic orbits for system (2.83) implies the GAS of $(i^*, r^*)^T$ in $\mathring{\mathcal{D}}_1$, and hence of $(s^*, i^*, r^*)^T$ in $\mathring{\mathcal{D}}$.

Actually for the case $\delta = 0$ we may still refer to the general structure discussed in Sect. 2.3. In fact if one considers system (2.83), it can be always reduced to the form (2.18) if we define $z = (i, r)^T$,

$$A = \begin{pmatrix} \epsilon - \lambda & -\lambda \\ \epsilon + c/r^* & 0 \end{pmatrix}, \quad e = \begin{pmatrix} \gamma(R_o - 1) \\ -(b + f) \end{pmatrix}$$

and

$$B = \begin{pmatrix} 0 & 0 \\ c & 0 \end{pmatrix}.$$

Suppose that a $z^* \in \mathring{\mathcal{D}}_1$ exists ($\mathring{\mathcal{D}}_1$ is invariant for our system), we may define $\tilde{A}$ as in (2.35) to obtain

$$\tilde{A} := A + diag(z^{*-1})B = \begin{pmatrix} \epsilon - \lambda & -\lambda \\ \epsilon + c/r^* & 0 \end{pmatrix}$$

which is sign skew-symmetric in the case $\epsilon < \lambda$. It is then possible to find a $W = diag(w_1, w_2)$, $w_i > 0$ such that $W\tilde{A}$ is essentially skew-symmetric; in fact its diagonal terms are nonpositive; we are in case (A) of Sect. 2.3. The associated graph is ○—●, and Theorem 2.3 applies to show GAS of z^* in $\mathring{\mathcal{D}}_1$.

Altogether it has been completely proven that, for this model too, the "classical" conjecture according to which a nontrivial endemic state z^* whenever it exists is GAS, still holds.

The same conjecture was made in [166] about an AIDS model with excess death rate of newborns due to vertical transmission.

We shall analyze this model in the next section.

As far as the behavior of $N(t)$ is concerned, the following lemma holds.

Lemma 2.9. [44] *Under the assumptions of Theorem 2.8, the total population $N(t)$ for system (2.71) has the following asymptotic behavior :*

a) *if $R_1 < 1$ then* $N(t) \downarrow 0$, *as* $t \to \infty$
   if $R_1 > 1$ then $N(t) \uparrow +\infty$, *as* $t \to \infty$
b) *the asymptotic rate of decrease or increase is $d(R_1 - 1)$ when $R_o < 1$, and the asymptotic rate of increase is $(d + \epsilon i^* + \delta r^*)(R_1 - 1)$ when $R_o > 1$.*

The behavior of $(S(t), I(t), R(t))$ is a consequence of the following lemma.

Lemma 2.10. [44] *The total number of infectives $I(t)$ for the model (2.71) decreases to zero if $R_2 < 1$ and increases to infinity if $R_2 > 1$. The asymptotic rate of decrease or increase is given by* $(c + d + \epsilon)(R_2 - 1)$.

The complete pattern of the asymptotic behavior of system (2.71) is given in Table 2.1 .

Table 2.1. Threshold criteria and asymptotic behavior [44]

R_o	R_1	R_2	$N \to$	$(s,i,r) \to$	$(S,I,R) \to$
≤ 1	< 1	$< 1^a$	0	$(1,0,0)$	$(0,0,0)$
> 1	< 1	$< 1^a$	0	(s^*,i^*,r^*)	$(0,0,0)$
≤ 1	> 1	< 1	∞	$(1,0,0)$	$(\infty,0,0)$
≤ 1	> 1	> 1	∞	$(1,0,0)$	(∞,∞,∞)
> 1	> 1	$> 1^a$	∞	(s^*,i^*,r^*)	(∞,∞,∞)

[a] Given R_o, R_1, this condition is automatically satisfied

2.3.2.4. An SIR model with vertical transmission and varying population. A model for AIDS [166]

A basic model to describe demographic consequences induced by an epidemic has been recently proposed by Anderson, May and McLean [166], in connection with the mathematical modelling of HIV/AIDS epidemics (see also Sect. 3.4).

With our notation, the model is based on the following set of ODE's

$$\begin{cases} \dfrac{dS}{dt} = b[N - (1-\alpha)I] - dS - \dfrac{\lambda}{N} IS \\ \dfrac{dI}{dt} = \dfrac{\lambda}{N} IS - (c + d + \epsilon)I \\ \dfrac{dR}{dt} = cI - dR \end{cases} \tag{2.86}$$

The total population $N(t)$ will then be a dynamical variable subject to the following evolution equation

$$\frac{dN}{dt} = b(N - (1-\alpha)I) - dN - \epsilon I \tag{2.87}$$

System (2.86) can be seen as a modification of system (2.71) with $e = \delta = 0$ (no loss of immunity after the disease, no excess death rate in the recovered class), and with a total birth rate reduced by the quantity $(1-\alpha)I$, $\alpha \in [0,1]$, due to vertical transmission of the disease; a fraction α of newborns from infected mothers may die at birth.

By introducing , as in Sect. 2.3.2.3, the fractions

$$s(t) = \frac{S(t)}{N(t)}\ , \quad i(t) = \frac{I(t)}{N(t)}\ , \quad r(t) = \frac{R(t)}{N(t)}, \qquad t \geq 0\ ,$$

we have that

$$s(t) + i(t) + r(t) = 1\ , \qquad t \geq 0 \tag{2.88}$$

so that we may reduce our analysis to the quantities $i(t)$, $r(t)$ in addition to $N(t)$.

The evolution equations for i and r are given by

$$\begin{cases} \dfrac{di}{dt} = -(b+c+\epsilon-\lambda)i - (\lambda-\epsilon-b(1-\alpha))i^2 - \lambda ir \\ \dfrac{dr}{dt} = -br + ci + (\epsilon + b(1-\alpha))ir \end{cases} \tag{2.89}$$

for $t > 0$, while the equation for N is given by

$$\frac{dN}{dt} = (b-d)N - [b(1-\alpha)+\epsilon]I\ , \qquad t > 0\ . \tag{2.90}$$

System (2.89) can be written in the form (2.18) if we define $z := (i, r)^T$, and

$$A = \begin{pmatrix} -\lambda+\epsilon+b(1-\alpha) & -\lambda \\ \epsilon+b(1-\alpha) & 0 \end{pmatrix}, \quad e = \begin{pmatrix} -(b+c+\epsilon-\lambda) \\ -b \end{pmatrix}$$

$$B = \begin{pmatrix} 0 & 0 \\ c & 0 \end{pmatrix}$$

The admissible space for system (2.89) is again

$$\mathcal{D}_1 := \left\{ (r,i)^T \in \mathbb{R}^2_+ \mid r+i \leq 1 \right\}.$$

The trivial solution $(r,i)^T = (0,0)^T$ always exists, and it is shown in [166] that a nontrivial endemic state $z^* \in \overset{\circ}{\mathcal{D}}_1$ exists for system (2.89) provided $R_o := \dfrac{\lambda}{c+\epsilon} > 1$ (again $\overset{\circ}{\mathcal{D}}_1$ is invariant for our system).

We may define $\tilde{A}$ as in (2.35) to obtain

$$\tilde{A} := A + diag(z^{*-1})B = \begin{pmatrix} -\lambda+\epsilon+b(1-\alpha) & -\lambda \\ \epsilon+b(1-\alpha)+c/r^* & 0 \end{pmatrix};$$

for $\lambda > \epsilon + b(1-\alpha)$ it is sign skew-symmetric. It is then possible to find a $W = diag(w_1, w_2)$, $w_i > 0$ such that $W\tilde{A}$ is essentially skew-symmetric; in fact its diagonal terms are nonpositive; we are in case (A) of Sect. 2.3. The associated graph is ○—●, and Theorem 2.3 applies to show GAS of z^* in $\overset{\circ}{\mathcal{D}}_1$.

As far as the asymptotic behavior of $N(t)$, and of the absolute values of $(S(t), I(t), R(t))$ we refer to Table 2.1 in Sect. 2.3.2.3.

We may like to point out that model (2.86) includes, for $\alpha = 1$, the case with no vertical transmission, and corresponds to the model proposed by Anderson and May [6] for host-microparasite associations (see also [163]).

On the other hand , for $\alpha = 0$, we have complete vertical transmission.

Other models of this kind have been considered in [35, 183]. In these papers the force of infection has a more general dependence upon the total population N, so that the transformation (2.73) does not eliminate the dependence upon N; specific analysis is needed in that case. As an example we have included an outline of the results obtained in [183] in Sect. 3.3. For the other cases we refer to the literature.

2.3.4. Multigroup models

A class of epidemic models of particular interest is related to the possibility that heterogeneous populations may participate to the epidemic process with different parameters. The case of spatially heterogeneous populations may be considered as part of this class whenever a (discrete) compartmental approach is allowed. Many authors have faced this problem to take into account socially structured populations [218]; to consider towns and villages grouping of a population [115]; to consider sexually transmitted diseases [118, 142] or other problems [114, 193, 208, 209, etc.]. But while for SIS type models a rather complete analysis of existence and global stability analysis of a nontrivial endemic state has been carried out by Lajmanovich and Yorke [142], by using techniques of monotonicity of the evolution operator (see Sect. 4.3.5), only partial results are known for SIR type models [208, 209, 29]. This is mainly due to the fact that SIR type models with vital dynamics do not show in general monotone behavior; on the contrary the trajectories spiral around the nontrivial endemic state, when this exists (see Figs. 2.1-2.3).

Anyway the SIS model for n dissimilar groups [142] can be seen as an extension of the model for two groups already discussed in Sect. 2.2.3 and Sect. 2.3.1.2.3. As such we will show that we can still put it in the general form (2.18).

2.3.4.1. SIS model for n dissimilar groups. A model for gonorrhea in an heterogeneous population [142]

The assumptions we shall make on the epidemic are the following :

(i) subpopulation $i \in \{1, \dots, n\}$ has a constant size $N_i \in \mathbb{N} - \{0\}$. The sizes of the susceptible class, and the infective class in the i-th population at time $t \geq 0$ will be denoted by $S_i(t), I_i(t)$, respectively. There is no migration among subpopulations; because of the infection process individuals are transferred among S_i and I_i classes in the i-th subpopulation but cannot be transferred to other subpopulations.

(ii) Births and deaths occur in the i-th subpopulation at equal rates for the two classes S_i, I_i and it is assumed that all newborns are susceptibles. The death process is assumed to be a linear decay with rate $\mu_i > 0$, the same for the whole i-th population, independent of the class S_i or I_i. This implies that there is no excess mortality induced by the disease. The constancy of the total size of the i-th population imposes the birth rate to be equal to μ_i.

(iii) Infective individuals of the j-th class ($j = 1, 2, \dots, n$) may transmit the disease to individuals of the i-th class ($i = 1, 2, \dots, n$); it is assumed that this occurs at a rate $\lambda_{ij} \geq 0\,(i, j = 1, \dots, n)$ according to the "law of mass action" so that the force of infection acting on the susceptibles S_i of the i-th group is given by

$$g_i(I_1, \ldots, I_n) = \sum_{j=1}^{n} \lambda_{ij} I_j\,, \quad i = 1, \ldots, n. \tag{2.91}$$

(iv) For the i-th group the recovery process is assumed to be a linear decay from the I_i class to the S_i class at a rate $\gamma_i > 0\,(i = 1, \ldots, n)$.

(v) All the parameters introduced above, $\mu_i, \lambda_{ij}, \gamma_i\,(i, j = 1, \ldots, n)$ are assumed to be time independent.

If we assume, for simplicity, that all subpopulation sizes $N_i = S_i(t) + I_i(t) = 1$, $i = 1, \ldots, n$, the above assumptions lead to the following family of ODE's :

$$\begin{cases} \dfrac{dI_i}{dt} = \left(\displaystyle\sum_{j=1}^{n} \lambda_{ij} I_j\right)(1 - I_i) - \gamma_i I_i - \mu_i I_i \\ I_i + S_i = 1 \end{cases} \tag{2.92}$$

for $i = 1, \ldots, n$, subject to suitable initial conditions.

It is easily seen that system (2.92) can be put in the form (2.18) if we define $z := (I_1, \ldots, I_n)^T$, and

$$A := -\Lambda\,, \quad \text{with} \quad \Lambda := (\lambda_{ij})_{i,j=1,\ldots,n} \tag{2.93}$$

$$e := (\lambda_{ii} - \gamma_i - \mu_i)_{i=1,\ldots,n} \tag{2.94}$$

$$B := \Lambda - diag\left[(\lambda_{ii})_{i=1,\ldots,n}\right] \tag{2.95}$$

For n much larger than 2 it is quite difficult to carry out the analysis of this system as we did in Sect. 2.3.1.2.3 for $n = 2$ for the technical reason of the difficulty of handling by paper and pencil large matrices. So while we can conjecture that condition (B) holds for the matrix $\tilde{A}$ defined by (2.35), it is preferable to adopt monotone techniques later in Sect. 4.3.5, to show existence and global asymptotic stability of a nontrivial endemic state for system (2.92).

2.3.4.2. SIR model for n dissimilar groups [29]

Such a model is obtained if we allow disease induced immunity so that individuals in the i-th group decay from the class I_i into a removed class R_i, $i = 1, \ldots, n$.

Thus now the total population N_i of the i-th group is divided into three subclasses of susceptibles S_i, of infectives I_i, and of immune (removed) R_i ;

$$N_i = S_i(t) + I_i(t) + R_i(t) \ , \qquad i = 1, \ldots, n \ .$$

With respect to the SIS model in Sect. 2.3.4.1 we only need to modify assumption (iv) which now becomes

(iv') for the i-th group the recovery to immunity is assumed to be a linear decay from the I_i class to the R_i class at a rate $\gamma_i > 0 (i = 1, \ldots, n)$.

If we assume for simplicity that all subpopulation sizes $N_i = S_i(t) + I_i(t) + R_i(t) = 1$, the above assumptions lead to the following family of ODE's:

$$(2.96) \qquad \begin{cases} \dfrac{d}{dt} S_i = -\left(\displaystyle\sum_{j=1}^{n} \lambda_{ij} I_j \right) S_i + \mu_i - \mu_i S_i \\ \dfrac{d}{dt} I_i = \left(\displaystyle\sum_{j=1}^{n} \lambda_{ij} I_j \right) S_i - (\mu_i + \gamma_i) I_i \\ \dfrac{d}{dt} R_i = \gamma_i I_i - \mu_i R_i \end{cases}$$

for $i = 1, \ldots, n$ subject to suitable initial conditions.

We shall see now that also system (2.96) can be set in the general form (2.27) with suitable transformations.

Since $S_i(t) + I_i(t) + R_i(t) = 1$, $i = 1, \ldots, n$ at any time $t \geq 0$, we may ignore the third equation in (2.96).

If we now introduce a new family of variables

$$(2.97) \qquad \chi_i(t) := S_i(t) + I_i(t) \, , \qquad t \geq 0$$

for $i = 1, \ldots, n$, system (2.96) can be rewritten as

$$(2.98)\qquad \begin{cases} \dfrac{d}{dt}S_i = \mu_i - \mu_i S_i + \left(\sum_{j=1}^{n} \lambda_{ij} S_j\right) S_i - \left(\sum_{j=1}^{n} \lambda_{ij} \chi_j\right) S_i \\ \dfrac{d}{dt}\chi_i = -(\mu_i + \gamma_i)\chi_i + (\mu_i + \gamma_i S_i) \end{cases}$$

for $i = 1, \ldots, n$. If we now denote

$$z := (S_1, \ldots, S_n, \chi_1, \ldots, \chi_n)^T \in \mathbb{R}^{2n}$$
$$c := (\mu_1, \ldots, \mu_n, \mu_1, \ldots, \mu_n)^T \in \mathbb{R}^{2n}$$
$$e := (-\mu_1, \ldots, -\mu_n, -(\mu_1 + \gamma_1), \ldots, -(\mu_n + \gamma_n))^T \in \mathbb{R}^{2n}$$

$$A := \left(\begin{array}{c|c} \Lambda & -\Lambda \\ \hline 0 & 0 \end{array}\right), \qquad \text{with} \quad \Lambda := (\lambda_{ij})_{i,j=1,\ldots,n}$$

$$B := \begin{pmatrix} 0 & 0 \\ \Gamma & 0 \end{pmatrix} \qquad \text{with} \quad \Gamma := diag(\gamma_1, \ldots, \gamma_n)$$

system (2.98) becomes

$$\frac{dz}{dt} = diag(z)\,(e + Az) + (c + Bz)$$

which corresponds to the general form (2.27).

We may then proceed as in Sect. 2.3.1 to prove existence and GAS of a nontrivial endemic state. By Corollary 2.4, since the vector c is strictly positive in $\mathbb{R}^{2n}$ then system (2.98) admits a strictly positive equilibrium $z^* \in \overset{\circ}{\Omega}$ where now $\Omega = \left\{ z \in \mathbb{R}_+^{2n} \mid \sum_{i=1}^{2n} z_i \leq 2n \right\}$.

The problem which is left open (for realistic values of the parameters) is to prove that $\tilde{A}$ defined in (2.35) satisfies either the conditions of Theorem 2.1 or those of Theorem 2.3 for GAS. The problem lies in the large dimensions of $\tilde{A}$, since we already know that for $n = 1$, or for non interacting populations ($\lambda_{ij} = 0$, for $i \neq j, i, j = 1, \ldots, n$) a unique nontrivial endemic state exists for each group (independently of each other) provided

$$(2.99)\qquad \lambda_{ii} > \mu_i + \gamma_i\,, \qquad i = 1, \ldots, n\ .$$

Figs. 2.1-2.3 [29] show by computer experiments the existence and GAS of a non trivial endemic state in a range of parameters which is more suitable for human diseases such as influenza, mumps, measles, etc. For this range of parameters $\mu \simeq 0$, as opposed to the other parameters, making the model rather insensitive to vital dynamics. In fact $I^* \simeq 0$ and the trajectories exhibit a behavior which corresponds more to the classical Kermack-McKendrick model.

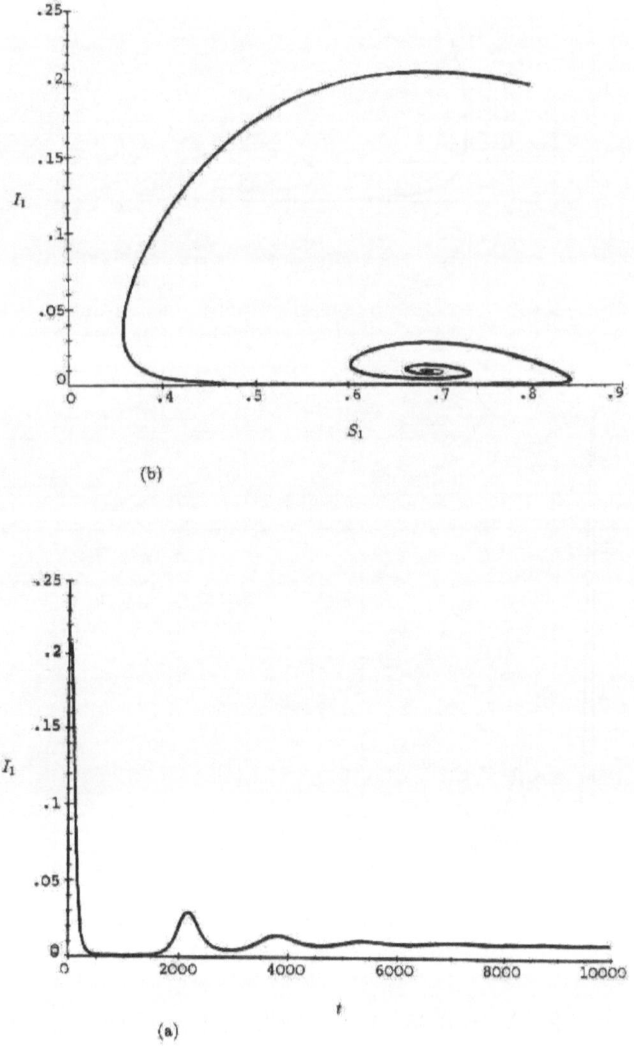

Fig. 2.1. $n = 2$; $\lambda_{ii} = 0.05$, $\lambda_{ij} = 0$, $i \neq j$, $i, j = 1, 2$; $\mu_i = 10^{-3}$, $\gamma_i = 30^{-1}$, $i = 1, 2$; $S_i^\circ = 0.8$, $I_i^\circ = 0.2$, $i = 1, 2$; [29].

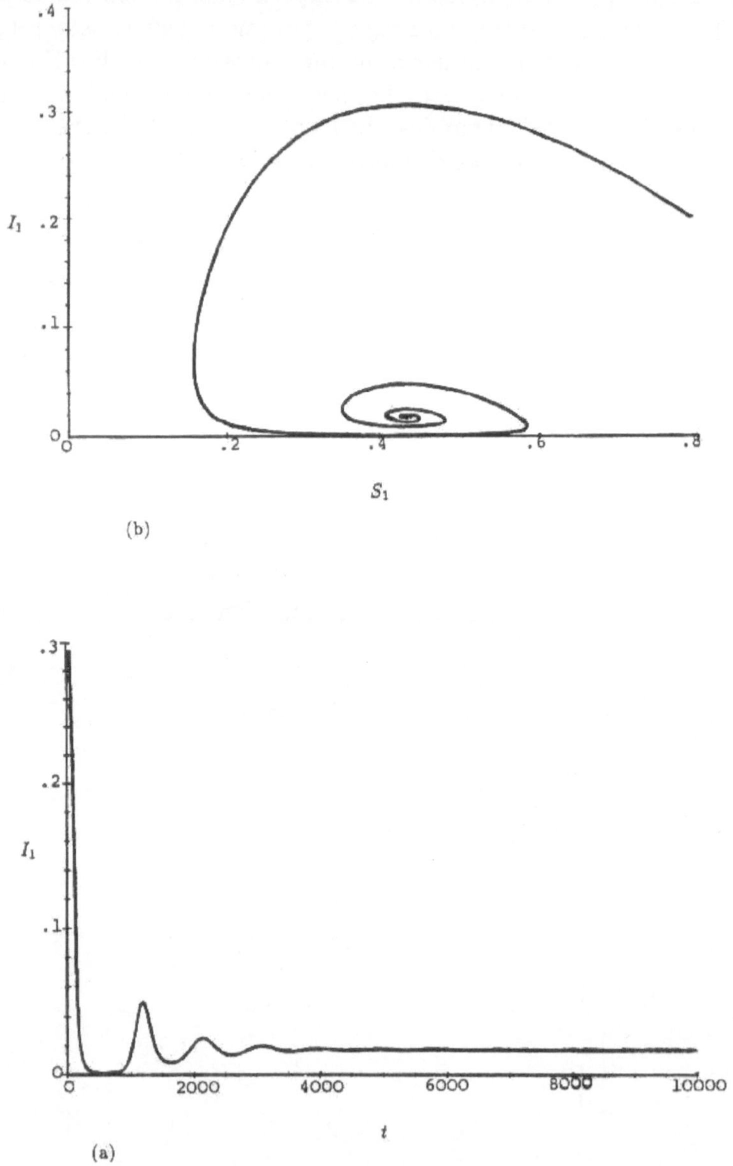

Fig. 2.2. $n = 2$; $\lambda_{ii} = 0.05$, $\lambda_{ij} = 0.03$, $i \neq j$, $i, j = 1, 2$; $\mu_i = 10^{-3}$, $\gamma_i = 30^{-1}$, $i = 1, 2$; $S_i^\circ = 0.8$, $I_i^\circ = 0.2$, $i = 1, 2$; [29].

Anyway "zooming" (Fig. 2.3) clearly show the spiraling of trajectories about the nontrivial endemic state that always exists.

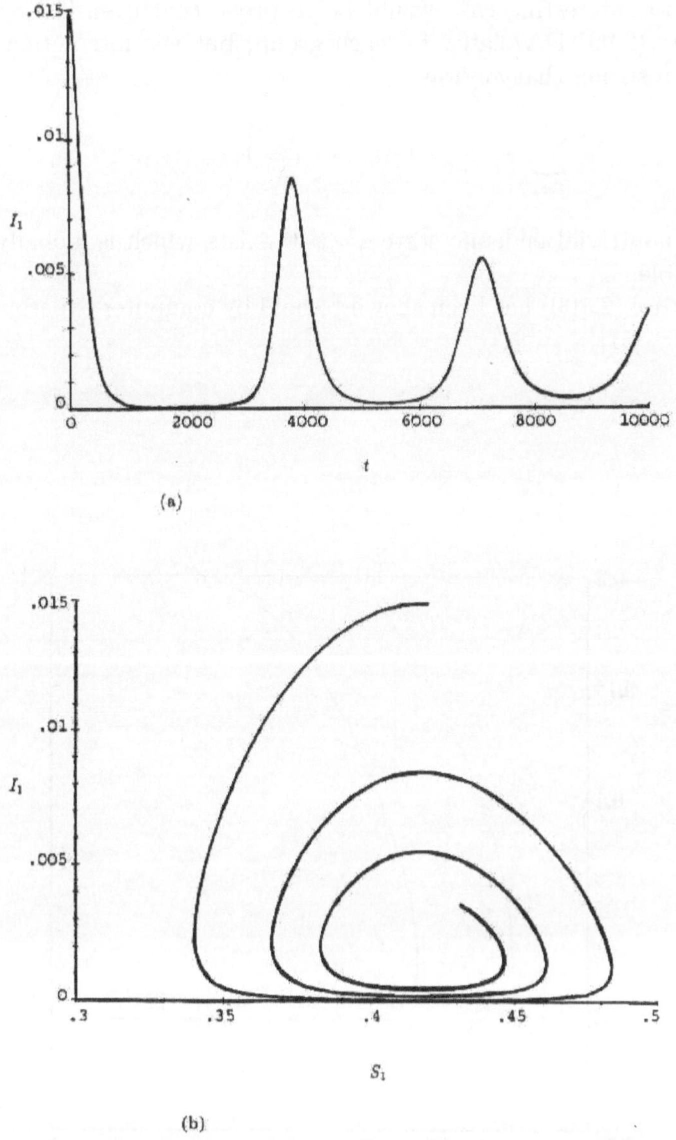

Fig. 2.3. $n = 2$; $\lambda_{ii} = 0.05$, $\lambda_{ij} = 0.03$, $i \neq j$, $i, j = 1, 2$; $\mu_i = 10^{-4}$, $\gamma_i = 30^{-1}$, $i = 1, 2$; $S_i^\circ = 0.8$, $I_i^\circ = 0.2$, $i = 1, 2$; [29].

Clearly, by continuous dependence of the solutions on the parameters, the same should hold for sufficiently small interactions $\lambda_{ij} \simeq 0$ for $i \neq j$ $(i, j = 1, \ldots, n)$, once (2.99) is satisfied.

The most interesting case would be to prove that even if $\lambda_{ii} = 0$, $i = 1, \ldots, n$, i.e. (2.99) is violated for each group, but the interaction between groups are so strong that anyway

$$\sum_{j=1}^{n} \lambda_{ij} > \mu_i + \gamma_i , \qquad i = 1, \ldots, n , \tag{2.100}$$

then still a nontrivial endemic state $z^* \gg 0$ exists which is globally asymptotically stable.

Conjecture (2.100) has been shown to hold by computer experiments (see Figs. 2.4-2.6 , [51]).

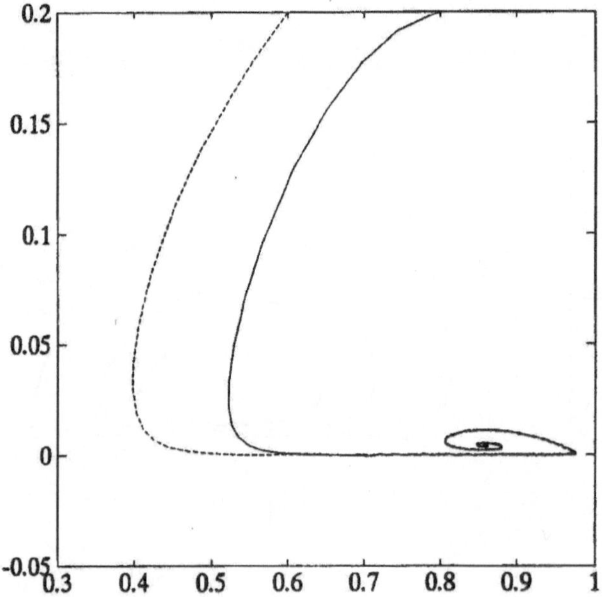

Fig. 2.4. $n = 2$; $\lambda_{ii} = 0$, $\lambda_{ij} = 0.04$, $i \neq j$, $i, j = 1, 2$; $\mu_i = 0.001$, $\gamma_i = 0.033$, $i = 1, 2$; $S_1^\circ = 0.8$, $S_2^\circ = 0.6$; $I_i^\circ = 0.2$, $i = 1, 2$; [51].

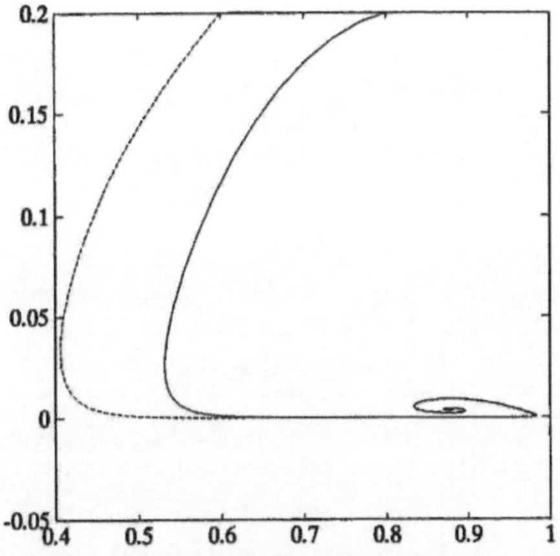

Fig. 2.5. $n=2$; $\lambda_{ii}=10^{-4}$, $\lambda_{ij}=0.039$, $i\neq j$, $i,j=1,2$; $\mu_i=0.001$, $\gamma_i=0.033$, $i=1,2$; $S_1^\circ=0.8$, $S_2^\circ=0.6$; $I_i^\circ=0.2$, $i=1,2$; [51].

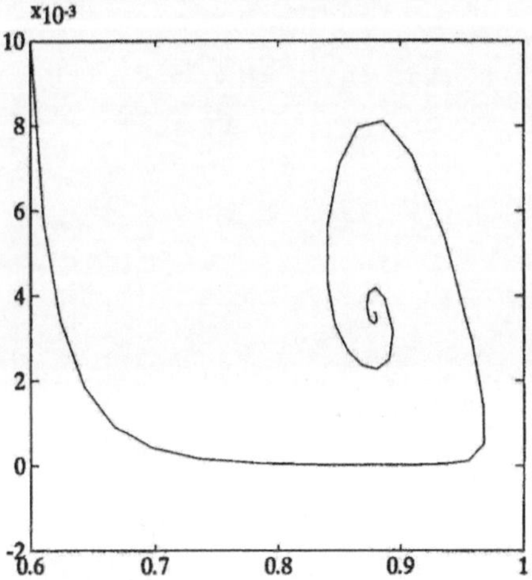

Fig. 2.6. $n=2$; $\lambda_{ii}=10^{-4}$, $\lambda_{ij}=0.039$, $i\neq j$, $i,j=1,2$; $\mu_i=0.001$, $\gamma_i=0.033$, $i=1,2$; $S_i^\circ=0.6$; $I_i^\circ=0.01$, $i=1,2$; [51].

3. Strongly nonlinear models

Up to now we have been considering epidemic models for which the rate of infection is bilinear in the infective (I) and in the susceptible (S) population; i.e. we have assumed that the force of infection due to the infective population is linear

$$g(I) = kI \tag{3.1}$$

corresponding to the classical "law of mass action". We have shown that, depending on the values of the parameters such models usually have a GAS trivial equilibrium corresponding to the disease free state, or a GAS nontrivial equilibrium corresponding to the endemic (persistent) state. In Sect. 2.3 and following we have shown that such a behavior is a direct consequence of a general theorem due to Beretta and Capasso [28] associated with the general structure of such epidemic systems. This structure includes in fact most of the well known "predator-prey like" or "symbiosis like" epidemic systems. When the restriction to linear force of infection and more generally to bilinear infection rate, is dropped, the system can have a much wider range of dynamical behaviors.

Wilson and Worcester were the first to consider the more general infection rate with a factor S^p . Severo [200] considers a more general form kI^pS^q, with $q < 1$, but did not motivate the behavior of the system in detail. Capasso and Serio [61] suggested that we have to consider the bilinear case as a special case of an infection rate of the form $g(I)S$ where $g(I)$, that we may call the force of infection, may be of any form, even if they analyzed only specific cases, by imposing $g'(0) \in (0, +\infty)$. Cunningham [74] pointed out that periodic solutions may appear in a model with infection rate $k(IS)^p$, provided $p > 1$.

More recently, Liu, Hethcote, Levin and Iwasa [155, 156] have faced the problem of a systematic analysis of the influence of nonlinear incidence rates of the form

$$h(I, S) = kI^pS^q \tag{3.2}$$

with $k, p, q > 0$.

The model they analyze is what is called an SEIRS model which includes, according to Table 3.1 all the basic Kermack-McKendrick type of "predator-prey like" models.

With respect to the classical SIRS model with vital dynamics the SEIRS models include a further class (E) of latent individuals which, having acquired the infection by the infectives, undergo a latent period, before being themselves capable of transmitting the disease.

Model	Equations	H	σ	Relations
SIS	$S' = -\lambda I^p S^q + \gamma I + \mu - \mu S$ $I' = \lambda I^p S^q - (\gamma + \mu) I$ $S + I = 1$	1	$\frac{k}{\gamma+\mu}$	limit of SIRS as $\delta \to \infty$ limit of SEIS as $\epsilon \to \infty$
SIR	$S' = -\lambda I^p S^q + \mu - \mu S$ $I' = \lambda I^p S^q - (\gamma + \mu) I$ $R' = \gamma I - \mu R$ $S + I + R = 1$	$\frac{\mu}{\gamma+\mu}$	$\frac{k}{\gamma+\mu}$	SIRS when $\delta = 0$
SIRS	$S' = -\lambda I^p S^q + \mu - \mu S + \delta R$ $I' = \lambda I^p S^q - (\gamma + \mu) I$ $R' = \gamma I - (\delta + \mu) R$ $S + I + R = 1$	$\frac{\delta+\mu}{\gamma+\delta+\mu}$	$\frac{k}{\gamma+\mu}$	limit of SEIRS as $\epsilon \to \infty$
SEIS	$S' = -\lambda I^p S^q + \mu - \mu S + \gamma I$ $E' = \lambda I^p S^q - (\epsilon + \mu) E$ $I' = \epsilon E - (\gamma + \mu) I$ $S + E + I = 1$	$\frac{\epsilon}{\gamma+\epsilon+\mu}$	$\frac{k\epsilon}{(\gamma+\mu)(\epsilon+\mu)}$	limit of SEIRS as $\delta \to \infty$
SEIR	$S' = -\lambda I^p S^q + \mu - \mu S$ $E' = \lambda I^p S^q - (\epsilon + \mu) E$ $I' = \epsilon E - (\gamma + \mu) I$ $R' = \gamma I - \mu R$ $S + E + I + R = 1$	$\frac{\epsilon\mu}{\gamma\epsilon+\mu(\epsilon+\gamma+\mu)}$	$\frac{k\epsilon}{(\gamma+\mu)(\epsilon+\mu)}$	SEIRS when $\delta = 0$
SEIRS	$S' = -\lambda I^p S^q + \mu - \mu S + \delta R$ $E' = \lambda I^p S^q - (\epsilon + \mu) E$ $I' = \epsilon E - (\gamma + \mu) I$ $R' = \gamma I - (\delta + \mu) R$ $S + E + I + R = 1$	$\frac{\epsilon(\delta+\mu)}{\gamma\epsilon+(\delta+\mu)(\epsilon+\gamma+\mu)}$	$\frac{k\epsilon}{(\gamma+\mu)(\epsilon+\mu)}$	

Table 3.1. Equations and Relations of Various Models [155]

3.1. The nonlinear SEIRS model

The SEIRS model can be written as follows

$$\text{(3.3)} \qquad \begin{cases} \dfrac{dS}{dt} = -kI^pS^q + \mu - \mu S + \delta R \\ \dfrac{dE}{dt} = kI^pS^q - (\epsilon + \mu)E \\ \dfrac{dI}{dt} = \epsilon E - (\gamma + \mu)I \\ \dfrac{dR}{dt} = \gamma I - (\delta + \mu)R \end{cases}$$

Under (3.3) we have $S + E + I + R = 1$.

As usual, we discuss the asymptotic behavior of system (3.3) for which the mathematical analysis allows us to state that the initial value problem is well posed in the sense that a unique solution exists for any choice of the initial data, for all $t \geq 0$, and depends continuously on the initial data and parameters.

The nonnegative orthant is positively invariant, for system (3.3); it always has a trivial equilibrium ($S = 1$; $E = I = R = 0$) which corresponds to the absence of the disease. Any equilibrium must satisfy

$$\text{(3.4)} \qquad E = \left(\frac{\gamma + \mu}{\epsilon}\right) I \; ; \qquad R = \frac{\gamma}{\delta + \mu} I \; .$$

As consequence, since all parameters are strictly positive, at any nontrivial equilibrium I, E and R are all strictly positive, as far as $S < 1$.

Any equilibrium must also satisfy

$$\text{(3.5)} \qquad kI^p (1 - E - I - R)^q = (\epsilon + \mu)E \; .$$

If we consider nontrivial equilibria, $(I > 0)$, then from (3.4) and (3.5) we get

$$\text{(3.6)} \qquad I^{p-1} \left(1 - \frac{I}{H}\right)^q = \frac{1}{\sigma}$$

where

$$H = \frac{\epsilon(\delta+\mu)}{\epsilon\gamma + (\epsilon+\gamma+\mu)(\delta+\mu)} \ ; \qquad \sigma = \frac{\epsilon k}{(\epsilon+\mu)(\gamma+\mu)} \tag{3.7}$$

The nontrivial equilibria (S^*, E^*, I^*, R^*) are such that

$$I^* < H$$

Once I^* is given, E^* and R^* are given by (3.4) and

$$S^* = 1 - \frac{I^*}{H} \ .$$

We look now for the number of nontrivial endemic states.
Denote by

$$f(I) := I^{p-1}\left(1 - \frac{I}{H}\right)^q \ . \tag{3.8}$$

The behavior of this function with different values of p and for any choice of $q > 0$ is shown in Fig. 3.1.

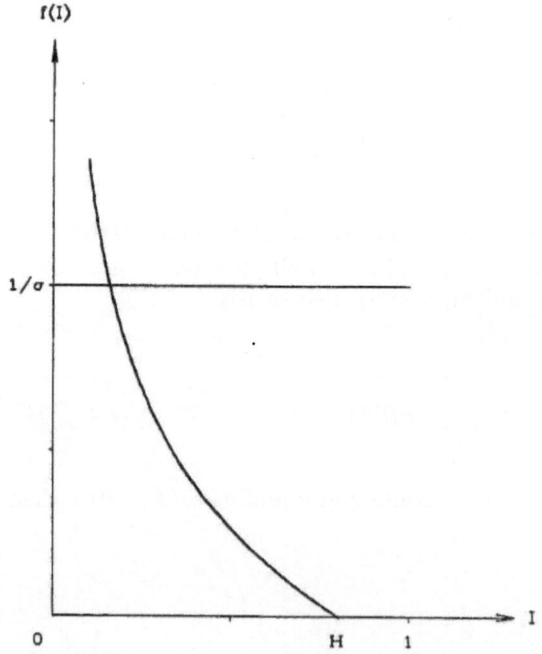

Fig. 3.1.a. $0 < p < 1$ [155]

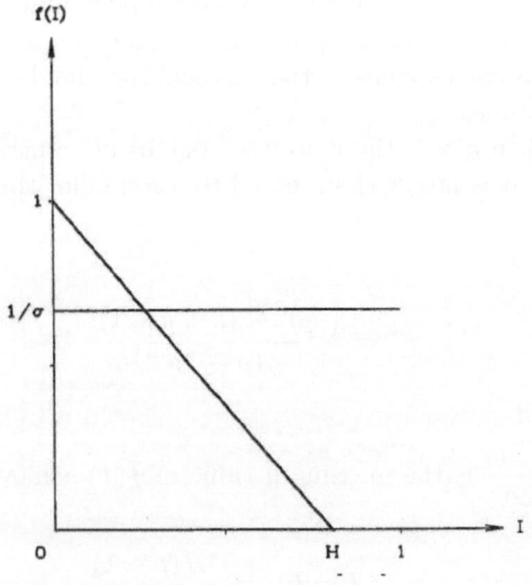

Fig. 3.1.b. $p = 1$ [155]

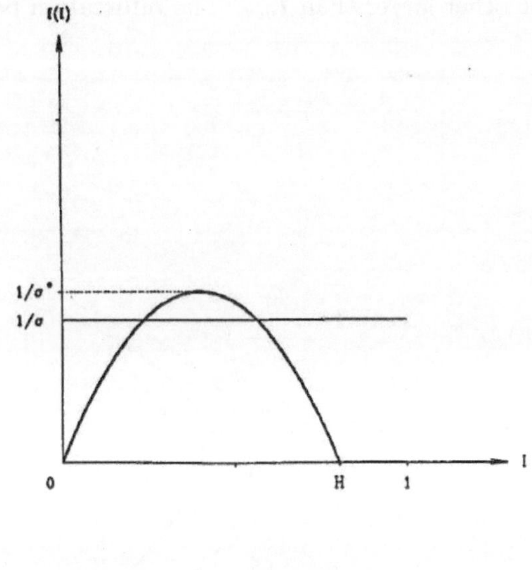

Fig. 3.1.c. $p > 1$ [155]

From Fig. 3.1, we can see that when $0 < p < 1$, there is always a unique nontrivial equilibrium. When $\sigma < 1$ and p tends to 1, the nontrivial equilibrium approaches the trivial equilibrium.

When $p = 1$, if $\sigma \leq 1$ there is no nontrivial equilibrium, whereas if $\sigma > 1$, there is a unique nontrivial equilibrium, which tends to the trivial one as σ tends to 1. This corresponds to the classical threshold theorem for SIR models with vital dynamics.

Finally when $p > 1$, there are two, one or no nontrivial equilibria according to whether σ is larger than, equal to, or smaller than σ^*, where

$$\sigma^* = \frac{\left(\frac{1}{H}\right)^{p-1}(p+q-1)^{p+q-1}}{(p-1)^{p-1}q^q} \tag{3.9}$$

Note that $\frac{1}{\sigma^*}$ is the maximum value of $f(I)$ achieved when

$$I = I_m := \frac{H(p-1)}{(p+q-1)}$$

and σ^* tends to 1 as p tends to one from above.

When $p > 1$ and $\sigma > \sigma^*$ the two nontrivial equilibria are one smaller than I_m, the other larger than I_m . The bifurcation pattern is shown in Fig. 3.2.

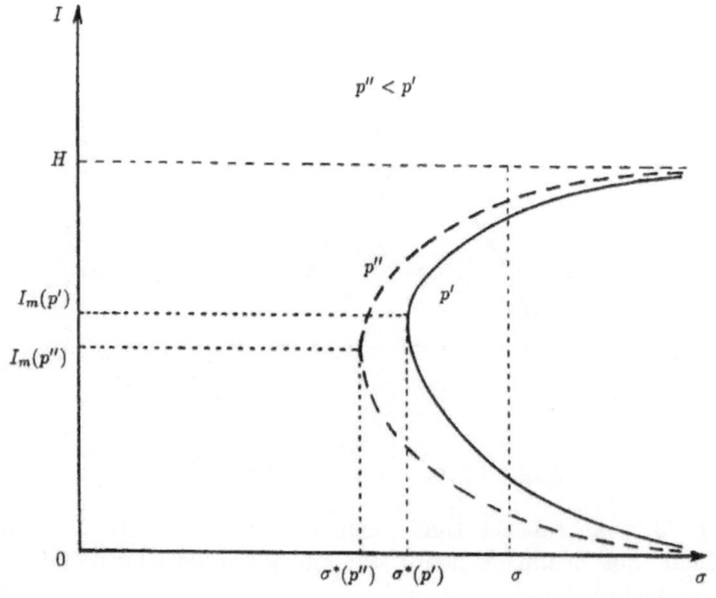

Fig. 3.2.

When $\sigma > \sigma^*$ and p tends to 1 from above, the small nontrivial equilibrium approaches the trivial equilibrium.

Figure 3.3 shows the behavior of the function $\sigma^*(p)$ and describes the number of equilibria in four regions of the $\sigma - p$ plane for the SEIRS model.

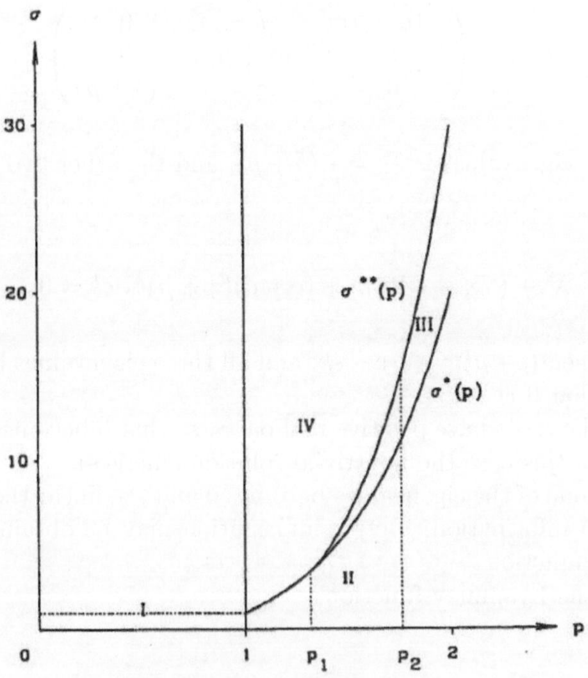

Fig. 3.3. [155]

Because of the assumption $S + E + I + R = 1$, the SEIRS model reduces to the 3-dimensional system

$$
(3.10) \qquad \begin{cases} \dfrac{dE}{dt} = kI^p(1 - E - I - R)^q - (\epsilon + \mu)E =: f_1(E, I, R) \\ \dfrac{dI}{dt} = \epsilon E - (\gamma + \mu)I \qquad\qquad =: f_2(E, I, R) \\ \dfrac{dR}{dt} = \gamma I - (\delta + \mu)R \qquad\qquad =: f_3(E, I, R) \end{cases}
$$

For $p > 1$, the Jacobi matrix at the trivial equilibrium $0 := (0, 0, 0)^T$ is

$$
J(0) = \begin{pmatrix} -(\epsilon + \mu) & 0 & 0 \\ \epsilon & -(\gamma + \mu) & 0 \\ 0 & \gamma & -(\delta + \mu) \end{pmatrix}
$$

Since $J(0)$ is a triangular matrix, its eigenvalues coincide with the diagonal elements, all of which are negative. Thus 0 is locally asymptotically stable (LAS), for $p > 1$.

For $p = 1$

$$J(0) = \begin{pmatrix} -(\epsilon+\mu) & k & 0 \\ \epsilon & -(\gamma+\mu) & 0 \\ 0 & \gamma & -(\delta+\mu) \end{pmatrix}$$

One of the eigenvalues is $\lambda_3 = -(\delta+\mu)$, and the other two are the roots of

$$\lambda^2 + (\gamma+\epsilon+2\mu)\lambda + (\epsilon+\mu)(\gamma+\mu) - \epsilon k = 0$$

If $\sigma < 1$, then $(\epsilon+\mu)(\gamma+\mu) > \epsilon k$, and all three eigenvalues have negative real parts, so that 0 is LAS.

If $\sigma > 1$, the roots have positive real parts, so that 0 becomes an unstable saddle point (in this case the nontrivial solution emerges).

For $\sigma = 1$ one of the eigenvalues becomes 0 and the linear theory does not provide enough information. Better information may be obtained by means of a Lyapunov function.

Consider the Lyapunov function :

$$V(z) = E + \frac{\epsilon+\mu}{\epsilon} I \tag{3.11}$$

where $z := (E, I, R)^T$. Its derivative along the trajectory of the system (3.3) is

$$\dot{V}(z) = \frac{(\epsilon+\mu)(\gamma+\mu)}{\epsilon} I \left[\sigma I^{p-1}(1-E-I-R)^q - 1\right] \tag{3.12}$$

For $p = 1$ and $\sigma \leq 1$, $\dot{V}(z) \leq 0$; the equality holds only for

(a) $\sigma = 1$ and $E = I = R = 0$

or

(b) $I = 0$

It is easily seen that the largest invariant subset of

$$R := \left\{ z \in \mathbb{R}^3 \mid \dot{V}(z) = 0 \right\}$$

reduces to 0.

Thus by the LaSalle Theorem [145] the trivial equilibrium is GAS in the invariant feasible region

$$\Omega := \left\{ z \in \mathbb{R}^3_+ \mid E + I + R \leq 1 \right\}$$

This analysis can be extended to the case $p > 1$.

Consider the quantity $\sigma I^{p-1}(1 - E - I - R)^q \leq \sigma I^{p-1}(1 - I)^q$.

The right hand side of the above inequality reaches its maximum $\dfrac{\sigma}{\sigma_1}$ when $I = (p-1)/(p+q-1)$, where $\sigma_1 := \dfrac{(p+q-1)^{p+q-1}}{(p-1)^{p-1}q^q}$.

Thus for $\sigma < \sigma_1$:

$$\sigma I^{p-1}(1 - E - I - R)^q \leq \frac{\sigma}{\sigma_1} < 1 \;, \tag{3.13}$$

and $\dot{V}(z) < 0$.

By the same arguments as above we may then claim that also for $p > 1$, 0 is GAS in Ω provided $\sigma < \sigma_1$.

When $p < 1$, in a region of Ω sufficiently close to 0, the quantity $(1 - E - I - R)^q$ is bounded away from zero.

Thus we can always find a value of I sufficiently small that $\dot{V}(z) > 0$.

Hence, all solutions but those starting on the R axis move away from the origin when $p < 1$.

3.1.1. Stability of the nontrivial equilibria

It can be shown that a nontrivial equilibrium z^* the Jacobi matrix is given by

$$J(z^*) = \begin{pmatrix} -(bHy + \epsilon + \mu) & bp - bHy & -bHy \\ \epsilon & -(\gamma + \mu) & 0 \\ 0 & \gamma & -(\delta + \mu) \end{pmatrix} \tag{3.14}$$

where

$$y := q\frac{I^*}{H}\left(1 - \frac{I^*}{H}\right)$$

and

$$b := \frac{k}{\sigma} = \frac{(\epsilon + \mu)(\gamma + \mu)}{\epsilon}$$

According to the Routh-Hurwitz criterion [27, 93], the necessary and sufficient conditions that all eigenvalues of J have negative real parts are

1) $tr := trace(J) < 0$

2) $\det := \det(J) < 0$

and

3) $C := tr\ M - det < 0$

where M is the sum of the second-order principal minors of J,

$$M := (\epsilon + \gamma + \delta + 2\mu)bHy + (\delta + \mu)(\epsilon + \gamma + 2\mu) - (\epsilon + \mu)(\gamma + \mu)(p - 1)$$

Now, the trace of J is given by

$$tr = -(\epsilon + \gamma + \delta + 3\mu + bHy)$$

is always a negative quantity. The determinant may be written as

$$det = (\epsilon + \mu)(\gamma + \mu)(\delta + \mu)(p - 1 - y) \ . \tag{3.15}$$

When $p \leq 1$, and the nontrivial equilibrium exists (i.e., $0 < p < 1$ or ($p = 1$ and $\sigma > 1$)), then $I^* < H$ so that $y > 0$, and det < 0. It can also be shown that in this case $C < 0$ [156].

Thus 1), 2), 3) are all satisfied and the nontrivial equilibrium is LAS.

When $p > 1$ and $\sigma > \sigma^*$, there is no nontrivial equilibrium; at $\sigma = \sigma^*$ there is a saddle-node bifurcation, and the unique nontrivial equilibrium is unstable.

When $p > 1$ and $\sigma > \sigma^*$, there are two nontrivial equilibria.

At the small nontrivial equilibrium $det > 0$, and at the large one $det < 0$. Therefore, the small nontrivial equilibrium is always an unstable saddle, and the stability of the large nontrivial equilibrium depends on the sign of C. When $C < 0$ the large nontrivial equilibrium is LAS ; when $C > 0$ it is unstable. When $C = 0$, an Hopf bifurcation may occur.

Tables 3.2 and 3.3 give a synthetic view of the different cases. For the analysis we refer to papers [155, 156].

To conclude we say with Liu et al. that although choosing a q value different from 1 has by itself no major effects, choosing a p value different from 1 can cause qualitative changes in the behavior of the system. For p large enough periodic solutions may occur. Actually the trivial solution coexists and is still LAS; therefore, depending on the initial conditions, the disease may die out or persist periodically. Thus the periodic orbit is not GAS.

Table 3.2. Equilibria and Their Stability for the SEIRS Model [155]

<table>
<tr><th colspan="2">Conditions</th><th>Stability of trivial equilibrium</th><th>Number of nontrivial equilibria</th><th colspan="3">Stability of nontrivial equlibria</th></tr>
<tr><td colspan="2">$0 < p < 1$</td><td>unstable saddle</td><td>1</td><td colspan="3">locally asymptotically stable</td></tr>
<tr><td rowspan="2">$p = 1$</td><td>$\sigma \leq 1$</td><td>globally asymptotically stable</td><td>0</td><td colspan="3"></td></tr>
<tr><td>$\sigma > 1$</td><td>unstable saddle</td><td>1</td><td colspan="3">locally asymptotically stable</td></tr>
<tr><td rowspan="5">$p > 1$</td><td>$\sigma < \sigma^*$</td><td>locally asymptotically stable[1]</td><td>0</td><td colspan="3"></td></tr>
<tr><td>$\sigma = \sigma^*$</td><td>locally asymptotically stable</td><td>1</td><td colspan="3">unstable</td></tr>
<tr><td rowspan="3">$\sigma > \sigma^*$</td><td rowspan="3">locally asymptotically stable</td><td rowspan="3">2</td><td rowspan="2">small nontrivial equilibrium</td><td colspan="2">large nontrivial equilibrium</td></tr>
<tr><td>$(\delta+\mu)^2 \geq \gamma\epsilon$</td><td>$(\delta+\mu)^2 < \gamma\epsilon$</td></tr>
<tr><td>unstable saddle</td><td>locally asymptotically stable</td><td>see Table 3.3</td></tr>
</table>

Fig. 3.3. [155]

[1] Global asymptotic stability is proved in Section 3.1 when $\sigma < \sigma_1$, where $\sigma_1 = \frac{(p+q-1)^{p+q-1}}{(p-1)^{p-1} q^q}$, and $\sigma^* = \sigma_1 / H^{p-1}$

Table 3.3.

$C < 0$	$C > 0$	$C = 0$
LAS	unstable	Hopf bifurcation may occur

3.2. A general nonlinear SEIRS model [116]

In Sect. 3.1 it was made clear via a systematic analysis of the influence of nonlinear incidence rates of the form

$$h(I,S) = kI^p S^q \tag{3.2}$$

with $k, p, q > 0$, that the dynamics is rather insensitive to the parameter q, while p is confirmed to be a relevant factor.

On the other hand, if we assume $q = 1$ in (3.2) we may include also this case in the framework of our presentation by saying that (3.2) corresponds (for $q = 1$) to a force of infection, acting on susceptibles, given by

$$g(I) = kI^p \tag{3.16}$$

with $p > 0$.

By revisiting the generalization of the Kermack-McKendrick model due to Capasso-Serio [61], in [116] Hethcote and van den Driessche propose the analysis of system (3.3) with a larger class of forces of infection $g(I)$ so that the nonlinear incidence rate is given by

$$h(I,S) = g(I)\, S \tag{3.16'}$$

where

$$\begin{aligned} &\text{(i)} \quad g(0) = 0 \\ &\text{(ii)} \quad g(I) > 0 \quad \text{for} \quad I > 0 \\ &\text{(iii)} \quad g(I) \in C^3(0,1)\ . \end{aligned} \tag{3.17}$$

With this in mind model (3.3) can be rewritten as

$$\begin{cases} \dfrac{dS}{dt} = -g(I)S + \mu - \mu S + \delta R \\ \dfrac{dE}{dt} = g(I)S - (\epsilon + \mu)E \\ \dfrac{dI}{dt} = \epsilon E - (\gamma + \mu)I \\ \dfrac{dR}{dt} = \gamma I - (\delta + \mu)R \end{cases} \tag{3.18}$$

so that the total population is constant

$$S(t) + E(t) + I(t) + R(t) = 1\,, \qquad t \geq 0\ . \tag{3.19}$$

Due to (3.19) system (3.18) can be reduced to

$$\begin{cases} \dfrac{dE}{dt} = g(I)\,(1 - E - I - R) - (\epsilon + \mu)E \\ \dfrac{dI}{dt} = \epsilon E - (\gamma + \mu)I \\ \dfrac{dR}{dt} = \gamma I - (\delta + \mu)R \end{cases} \tag{3.20}$$

subject to suitable initial conditions.

It is easily seen that the region

$$\Omega := \{(E, I, R) \in \mathbb{R}^3 \mid E \geq 0\,,\ I \geq 0\,,\ R \geq 0\,,\ E + I + R \leq 1\} \tag{3.21}$$

is positively invariant for our system. Conditions (3.17) imply that the initial value problem for (3.20) is well posed in Ω.

Due to (3.17)(i), the system always has the trivial, disease free equilibrium $(E, I, R) = (0, 0, 0)$ with $S = 1$. Any equilibrium must satisfy

$$E = \frac{\gamma + \mu}{\epsilon} I \quad ; \quad R = \frac{\gamma}{\delta + \mu} I \tag{3.22}$$

so that, if we assume that all parameters $\gamma, \mu, \epsilon, \delta > 0$, $I > 0$ implies $E > 0$ and $R > 0$.

Any nontrivial equilibrium must satisfy the equation

$$\frac{g(I)}{I}\left(1 - \frac{I}{H}\right) = \frac{1}{\sigma} \tag{3.23}$$

where

$$H := \frac{\epsilon(\delta + \mu)}{\gamma\epsilon + (\delta + \mu)\,(\epsilon + \gamma + \mu)} < 1\ ; \quad \sigma = \frac{\epsilon}{(\epsilon + \mu)\,(\gamma + \mu)} \tag{3.24}$$

which clearly corresponds to (3.6), (3.7) respectively.

Equation (3.23) shows that if we denote by (E^*, I^*, R^*) the nontrivial equilibrium, then again

$$I^* < H \tag{3.25}$$

Once I^* is given, E^* and R^* are given by (3.22); thus we only need to look for nontrivial solutions of (3.23) that we may rewrite in the form

$$f(I) = \frac{1}{\sigma} \tag{3.26}$$

once we define

$$f(I) := \frac{g(I)}{I}\left(1 - \frac{I}{H}\right) \qquad I \in (0, H] \ . \tag{3.27}$$

Note that

$$f(I) > 0 \quad \text{for} \quad I \in (0, H)$$

$$f(H) = 0$$

$$f(0+) = \lim_{I \to 0} \frac{g(I)}{I} \ .$$

The shape of $f(I)$, and better of $g(I)$ is crucial to solve (3.26).

Let us consider the slope of f in $(0, H)$:

$$f'(I) = \frac{g'(I)}{I}\left(1 - \frac{I}{H}\right) - \frac{g(I)}{I^2} \tag{3.28}$$

We shall analyze the cases

(i) $g(I) = kI$, law of mass action

(ii) $g(I) = kI^p$, $p > 0$ [155, 156]

(iii) $g(I) = \dfrac{kI^p}{1 + \alpha I^q}$, $p, q > 0$ [58, 61, 116]

The cases (i), (ii) have already been discussed in Sect. 3.1 (see Figs. 3.1 (a) -(c)) [155, 156].

We shall then discuss the case (iii) which includes for $p = q = 1$ a concave, saturating force of infection [61]; for $\alpha = 0$ the case (ii) [155, 156]; for $p = q = 2$ a sigma type saturating force of infection [58] (see also the discussion in Sect. 4.4).

For the case $p = q$ we have the behaviors illustrated in Fig. 3.4.

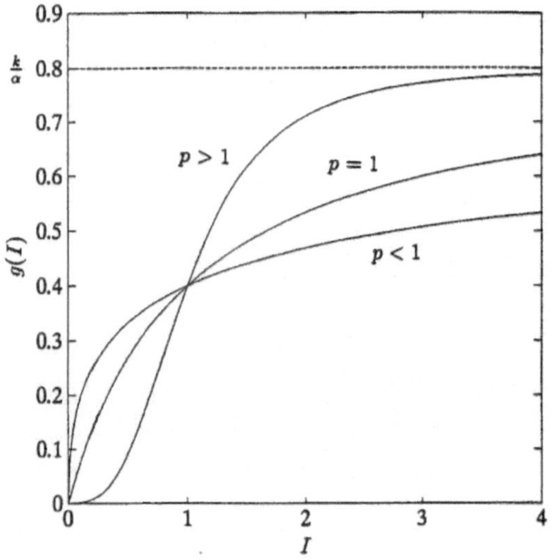

Fig. 3.4. $g(I) = k\frac{I^p}{1+\alpha I^p}$

For the case (iii) Eqn. (3.27) becomes

$$f(I) = k\frac{I^{p-1}}{(1+\alpha I^q)}\left(1-\frac{I}{H}\right) \tag{3.29}$$

and Eqn. (3.28) becomes

$$f'(I) = k\frac{I^{p-2}}{(1+\alpha I^q)^2}\left[(p+\alpha(p-q)I^q)\left(1-\frac{I}{H}\right)-(1+\alpha I^q)\right] \tag{3.30}$$

so that $f'(I) >=< 0$ depending upon the inequality

$$(p + \alpha(p-q)I^q)\left(1 - \frac{I}{H}\right) >=< (1 + \alpha I^q)$$

for the simplest case $p = q$ it becomes

$$p\left(1 - \frac{I}{H}\right) >=< (1 + \alpha I^q)$$

Depending upon the value of p as compared to 1 we may have, for $p > 1$ a unique point $I_m \in (0, H)$ at which $f'(I_m) = 0$, and for $I < I_m : f'(I) > 0$; for $I > I_m : f'(I) < 0$. Thus I_m is a maximum for f determined by

$$\frac{1}{\sigma^*} = f(I_m). \tag{3.31}$$

Hence depending upon the relative value of σ with respect to σ^* , we may have 0, 1, 2 non trivial solutions for our system.

A complete pattern of different cases is given in Table 3.4 [116].

Table 3.4. Number of nontrivial equilibria $\in \{0, 1, 2\}$ with $g(I) = I^p/(1 + \alpha I^q)$ for the two cases $q = p$ and $q = p - 1$ showing dependence on σ. [116]

p	$q = p$		$q = p - 1$	
< 1	all σ	1	$\sigma \leq \alpha$	0
			$\sigma > \alpha$	1
$= 1$	$\sigma \leq 1$	0	$\sigma \leq 1 + \alpha$	0
	$\sigma > 1$	1	$\sigma > 1 + \alpha$	1
> 1	$\sigma < \sigma^*$	0		
	$\sigma = \sigma^*$	1		
	$\sigma > \sigma^*$	2		

These facts are extended to the general case (3.16'), by means of the following theorem [116] which includes all cases (i)-(iii).

Theorem 3.1. *The system (3.20) always has the disease free equilibrium* $(0,0,0)$, *and if*

(i) $f(0) \leq 1/\sigma$ *and* $f'(I) < 0$ *for all* $I \in (0,H)$, *then it has no nontrivial equilibrium,*
(ii) $f(0) > 1/\sigma$ *and* $f'(I) < 0$ *for all* $I \in (0,H)$, *then it has 1 nontrivial equilibrium,*
(iii) $f(0) = 0$ *and* $\sigma < \sigma^*$, *then it has no nontrivial equilibrium,*
(iv) $f(0) = 0$ *and* $\sigma = \sigma^*$, *then it has 1 nontrivial equilibrium,*
(v) $f(0) = 0$ *and* $\sigma > \sigma^*$, *then it has 2 nontrivial equilibria* I_1, I_2 *where* $0 < I_1 < I_m < I_2 < H$.

3.2.1. Stability of equilibria

By the same techniques used in [155, 156] (see Sect.3.1) for the stability of the trivial equilibrium the results listed in Table 3.5 are obtained in [116], for the function g given in (iii).

Table 3.5. Stability of the disease free equilibrium with $g(I) = I^p/(1+\alpha I^q)$ for the two cases $q = p$ and $q = p - 1$ showing dependence on σ . [116]

p	$q=p$		$q=p-1$	
<1	unstable		$\sigma<\alpha$	GAS
			$\sigma>\alpha$	unstable
$=1$	$\sigma\leq 1$	GAS	$\sigma\leq 1+\alpha$	GAS
	$\sigma>1$	unstable	$\sigma>1+\alpha$	unstable
>1	LAS GAS for $\sigma<\sigma_1$			

Note: $\sigma_1 = \dfrac{1}{max\ \frac{g(I)}{I}\,(1-I)}$

The stability analysis of the nontrivial equilibria can be also carried out along the same lines as in [155, 156], with suitable modifications. The results obtained in [116] are listed in Table 3.6 . More details can be found in [116].

Table 3.6. Stability pattern with $g(I) = kI^p/(1+\alpha I^p)$

<table>
<tr><th>Conditions</th><th colspan="2">Stability of trivial equilibria</th><th>Number of nontrivial equilibria</th><th colspan="2">Stability of nontrivial equlibria</th></tr>
<tr><td>$0 < p < 1$</td><td colspan="2">unstable</td><td>1</td><td colspan="2">LAS</td></tr>
<tr><td rowspan="2">$p = 1$</td><td>$\sigma \leq 1$</td><td>GAS</td><td>0</td><td colspan="2"></td></tr>
<tr><td>$\sigma > 1$</td><td>unstable</td><td>1</td><td colspan="2">LAS</td></tr>
<tr><td rowspan="4">$p > 1$</td><td>$\sigma < \sigma^*$</td><td rowspan="4">LAS and GAS for $\sigma < \sigma_1$</td><td>0</td><td colspan="2"></td></tr>
<tr><td>$\sigma = \sigma^*$</td><td>1</td><td colspan="2">unstable</td></tr>
<tr><td rowspan="2">$\sigma > \sigma^*$</td><td rowspan="2">2</td><td>small equilibria</td><td>large equilibria</td></tr>
<tr><td>unstable saddle</td><td>see Tab. 3.3</td></tr>
</table>

Note: $\sigma_1 = \dfrac{1}{max\, \frac{g(I)}{I}(1-I)}$

3.3. An epidemic model with nonlinear dependence upon the population size

In Sect. 2.3.2 epidemic models with varying population size have been considered in which the dependence of the effective contact rate, and hence of the force of infection, upon the population size $N(t)$ has been taken into account (see system (2.71))

$$g(I;N) = \frac{\lambda}{N} I . \tag{3.32}$$

In such a case the incidence rate becomes

$$g(I;N)\, S = \frac{\lambda}{N} IS \ . \tag{3.33}$$

In recent papers [35, 36, 125, 182, 183] a more general dependence of the force of infection upon the total population has been suggested, as an extension of (3.32), as follows

$$g(I;N) = \sigma(N)\, I \tag{3.34}$$

where $\sigma : \mathbb{R}_+ \to \mathbb{R}_+$ is a general nonincreasing function, such that

(i) the function $c : N \in \mathbb{R}_+ \to N\sigma(N) \in \mathbb{R}_+$ is strictly increasing

(ii) $c \in C^1(\mathbb{R}_+)$

(iii) $\lim_{N \to 0+} c(N) = 0.$

In this case the incidence rate will be

$$g(I;N)\, S = \sigma(N)\, IS = \frac{c(N)}{N} IS \ . \tag{3.35}$$

On the other hand the intrinsic birth and death rates may be assumed themselves to be nonlinear functions of the total population, respectively given by $B(N)$ and $D(N)$.

One may assume that all newborns are in the susceptible class, if vertical transmission is ignored, and that the total population has a carrying capacity $K > 0$ so that [35, 36]

$$
\begin{aligned}
&B(K) = D(K) \\
&B'(K) < D'(K) \\
&B(N) \le D(N) \qquad \text{if} \qquad N \ge K
\end{aligned}
\tag{3.36}
$$

With all this in mind, we discuss here an SEI epidemic model proposed by Pugliese [183] to illustrate the mathematical techniques that can be used to solve such a problem, that clearly cannot be included in the general framework of Sect. 2.

For a more extensive treatment we refer to the literature.

The model is described by the following equations

$$
\begin{cases}
\dfrac{dS}{dt} = B(N)\,S + B(N)\,(1-\delta_1)\,E + B(N)\,(1-\delta_2)\,I \\
\qquad - D(N)\,S - \sigma(N)\,IS \\
\dfrac{dE}{dt} = \sigma(N)\,IS - D(N)\,E - \nu E \\
\dfrac{dI}{dt} = \nu E - (D(N)+\mu)\,I
\end{cases}
\tag{3.37}
$$

where $N = S + E + I$.

The specific features of this model are first, the differentiated fertility from individuals in classes E and I, which is obtained by multiplying the intrinsic birth rate $B(N)$ by the quantities $(1-\delta_1)$, and $(1-\delta_2)$ respectively; second, the differential mortality rate μ of infectives.

To simplify the analysis in the sequel it is assumed that $B(N) = a$ (a constant), and that either $\delta_1 = 0$ or $\delta_1 = \delta_2 = 1$.

More specifically it will be assumed that

(H1) $a > 0,\ \nu > 0,\ \mu \ge 0,\ 0 \le \delta_2 \le 1,\ \mu + a\delta_2 > 0$

(H2) an interval $(N_1, N_2) \subset (0, +\infty)$ exists such that $D(N)$ is strictly increasing on (N_1, N_2) and $D(N_1) < a < D(N_2)$.

As a consequence an $N^* > 0$ exists such that $D(N^*) = a$. This is the demographic equilibrium, in absence of disease.

The evolution equation for the total population is

$$\frac{dN}{dt} = (a - D(N))\,N - a\delta_1 E - a\delta_2 I - \mu I \; . \tag{3.38}$$

Since at any time $t \geq 0$, we have

$$N(t) = S(t) + E(t) + I(t) \tag{3.39}$$

to solve system (3.37) we can use alternatively the evolution system for (N, E, I).

In our case we get

$$\begin{cases} \dfrac{dN}{dt} = (a - D(N))N - a\delta_1 E - (a\delta_2 + \mu)I \\ \dfrac{dE}{dt} = \sigma(N)\,IN - \sigma(N)\,IE - \sigma(N)\,I^2 - D(N)\,E - \nu E \\ \dfrac{dI}{dt} = \nu E - (D(N) + \mu)\,I \end{cases} \tag{3.40}$$

System (3.40) always admits the trivial solution $(N^*, 0, 0)$, the stability properties of which are given in the following proposition.

Proposition 3.2. *Let*

$$R_o = \frac{\sigma(N^*)\,N^*}{D(N^*) + \mu}\,\frac{\nu}{a + \nu} \tag{3.41}$$

a) If $R_o < 1$, the trivial equilibrium $(N^*, 0, 0)$ is LAS for system (3.37).
b) If $R_o > 1$, $(N^*, 0, 0)$ is unstable

Proof. The Jacobi matrix of system (3.40) at $(N^*, 0, 0)$ is

$$J(N^*, 0, 0) = \begin{pmatrix} -D'(N^*)\,N^* & -a\delta_1 & -(a\delta_2 + \mu) \\ 0 & -(D(N^*) + \nu) & \sigma(N^*)\,N^* \\ 0 & \nu & -(D(N^*) + \mu) \end{pmatrix} \tag{3.42}$$

whose eigenvalues are given by

$$\lambda_1 = -D'(N^*)\,N^* < 0 \tag{3.43}$$

and the eigenvalues of

$$B = \begin{pmatrix} -(D(N^*)+\nu) & \sigma(N^*)\,N^* \\ \nu & -(D(N^*)+\mu) \end{pmatrix} \tag{3.44}$$

Clearly we have $tr B < 0$, while

$$\det B = (a+\nu)\,(a+\mu) - \nu\,\sigma(N^*)\,N^*. \tag{3.45}$$

The Routh-Hurwitz criterion implies a) and b).

As far as nontrivial endemic equilibria $\overline{z} := (\overline{N}, \overline{E}, \overline{I},)^T$, with $\overline{N}, \overline{E}, \overline{I} > 0$, are concerned, after some technical manipulations, it is easily seen that the following two propositions hold for the two cases $\delta_1 = 0$, and $\delta_1 = \delta_2 = 1$ respectively [183].

Proposition 3.3. *Let $\delta_1 = 0$, and let R_o be defined as in (3.41)*

a) If $R_o \leq 1$, then there are no nontrivial equilibria.
b) If $R_o > 1$, there exists a unique positive equilibrium.

Proposition 3.4. *Let $\delta_1 = \delta_2 = 1$, and let R_o be defined as in (3.41)*

a) If $R_o \leq 1$, then there may or may not exist nontrivial equilibria.
b) If $R_o > 1$, then there exists at least one nontrivial equilibrium.

As one can see from Proposition 3.4, it is not possible to get a general answer to the problem of existence of nontrivial endemic states. The question of the stability of these states is still more intricate. In [183] numerical results are given for specific choices of $D(N)$ and $\sigma(N)$.

The case in which the force of infection is given by (3.32), i.e. $\sigma(N) = \lambda/N$, or $c(N) = \lambda$, can be handled by the same techniques discussed in Sect. 2.3.2.3 and Sect. 2.3.2.4 [183, 44].

3.4. Mathematical models for HIV/AIDS infections

A large amount of literature is now available about the mathematical modelling of HIV/AIDS epidemics [9].

Most of these models, based on the classical mathematical epidemiology are characterized by specific modifications in order to take into account relevant peculiarities of the mechanisms of transmission of this infection.

Among those we may mention the substantial variability of the infection rates for different subpopulations at risk (homosexuals, heterosexuals, drug users, etc.); the long incubation period before the exhibition of symptoms; the variability of the infection with respect to the evolution of the infection in each individual; etc.

Moreover the above peculiar aspects of HIV/AIDS infections imply a relevant impact on the demography of the population at risk.

As a consequence the mathematical modelling of the dynamics of HIV/AIDS infections cannot avoid including "structures" in the population.

In order to avoid an high level of complexity most of the authors have proposed to start from simplified models in order to obtain some first information about the macroscopic qualitative behavior of the system. Later more realistic features have been introduced for a closer agreement with experimental data.

In the following we provide a first introduction to the subject, leading the interested reader to the existing literature.

3.4.1. One population models. One stage of infection

In Sect. 2.3.2 and Sect. 3.3 epidemic models with varying population size and with a nonlinear dependence of the force of infection upon the population size itself have been treated. In Sect. 2.3.2 the simplification came out of the specific choice of the dependence of the force of infection upon the total population (see also (3.32)). In both cases the "law of mass action" was somehow preserved since the force of infection due to the infected population I was given by

$$g(I;N) = \sigma(N)\,I \tag{3.46}$$

being N the total population (which in this case includes the population of removed individuals)

$$N = S + I + R\ .$$

In HIV/AIDS modelling, in contrast with what we have been discussing in Sect. 2.3.2.4, we have to take into account that removed individuals can be considered only those affected by AIDS symptoms and hence are not any more sexually active so that they cannot participate to the infection dynamics. This fact implies that in an SIR type model, the force of infection be given by [65, 125, 165]

$$g(I;T) = C(T)\, I\ . \tag{3.47}$$

where now

$$T = S + I\ . \tag{3.48}$$

The incidence rate will then be given by $g(I;T)\, S$.

Models of this kind include the one population model and a single stage of infection due to Jacquez et al. [125], and the one population model due to Castillo-Chavez et al. [65].

We start with the model in [125] for simplicity.

3.4.1.1. Model I [125]

We shall denote as usual by $S(t)$ the total number of susceptibles, by $I(t)$ the total of HIV infectives, and by $A(t)$ the total number of individuals who have fully developed AIDS, and thus they no longer take part in the transmission.

The force of infection is then assumed to be given by

$$g(I;T) = \frac{\lambda}{T}\, I \tag{3.49}$$

The mathematical model of SIR type with a constant input Λ of susceptibles and an intrinsic death rate μ, is then given by [125]

$$\begin{cases} \dfrac{dS}{dt} = -g(I;T)\, S - \mu S + \Lambda \\ \dfrac{dI}{dt} = g(I;T)\, S - (k+\mu) I \qquad , \ t > 0 \\ \dfrac{dA}{dt} = kI - (\mu + \delta) A \end{cases} \tag{3.50}$$

It has been assumed that all newborns are susceptibles and that individuals which have fully developed AIDS have a differential mortality rate δ.

The total population $N(t) = S(t) + I(t) + A(t)$ for system (3.50) evolves according to

$$\frac{dN}{dt} = -\mu N(t) - \delta A + \Lambda\,, \qquad t > 0 \tag{3.51}$$

Due to (3.51), system (3.50) satisfies the following properties .

Proposition 3.5. *For $\mu > 0$, if $N(0) \leq \Lambda/\mu$, then $N(t) \leq \Lambda/\mu$, for any $t \geq 0$. In particular the set*

$$\Omega := \left\{ (S, I, A) \in \mathbb{R}^3_+ \mid S + I + A \leq \frac{\Lambda}{\mu} \right\} \tag{3.52}$$

is a compact, convex, invariant set for system (3.50).

Proof. In fact from (3.51) it is clear that

$$\frac{dN}{dt} \leq -\mu N(t) + \Lambda$$

i.e.

$$\frac{dN}{dt} + \mu N(t) \leq \Lambda$$

Multiply by $e^{\mu t}$ and integrate from 0 to T, $T \geq 0$ to get

$$\left[N(T) - \frac{\Lambda}{\mu} \right] e^{\mu T} \leq N(0) - \frac{\Lambda}{\mu} \leq 0\ .$$

To show the invariance of Ω, we must show that solutions that start on the boundaries of Ω move into Ω, and this is a trivial task.

Proposition 3.6. *$N(t)$ is a dynamical variable. If $\delta = 0$ in the third equation (3.50) then*

$$N(t) = \left(N(0) - \frac{\Lambda}{\mu} \right) e^{-\mu t} + \frac{\Lambda}{\mu}\ ,$$

and

$$\lim_{t \to +\infty} N(t) = \frac{\Lambda}{\mu} \; .$$

Proposition 3.7. *For $\mu > 0$ system (3.50) admits the disease-free equilibrium*

$$S^* = \frac{\Lambda}{\mu}, \quad I^* = 0, \quad A^* = 0 \; . \tag{3.53}$$

As far as nontrivial equilibria are concerned we proceed as follows.

Since the first two equations in (3.50) are independent of A, we can restrict the analysis to these two, without loss of generality.

For $\mu > 0$, in the phase plane (S, I) the isocline $\frac{dS}{dt} = 0$ is given by the curves

$$-(\lambda + \mu)SI - \mu S^2 + \Lambda(S + I) = 0 \tag{3.54}$$

while the isocline $\frac{dI}{dt} = 0$ is given by the lines

$$I = 0 \qquad \text{and} \qquad I = \frac{\lambda - (k + \mu)}{k + \mu} S \tag{3.55}$$

We have then two phase portraits according to the sign of $\lambda - (k + \mu)$ (see Figs. 3.5 and 3.6).

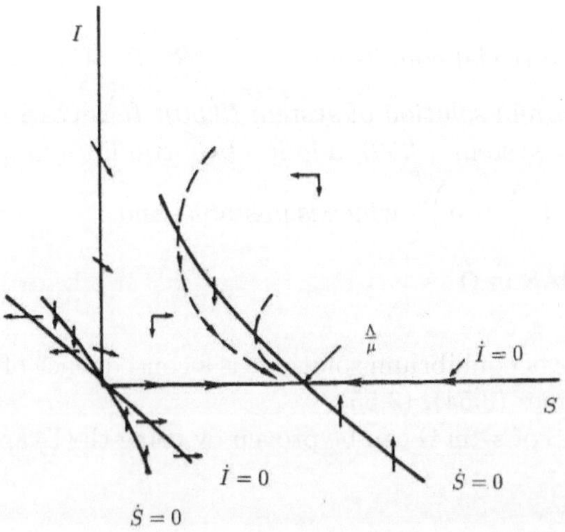

Fig. 3.5. Phase diagram for system (3.50) for $\mu > 0$ and $\lambda - (k+\mu) < 0$ [125]

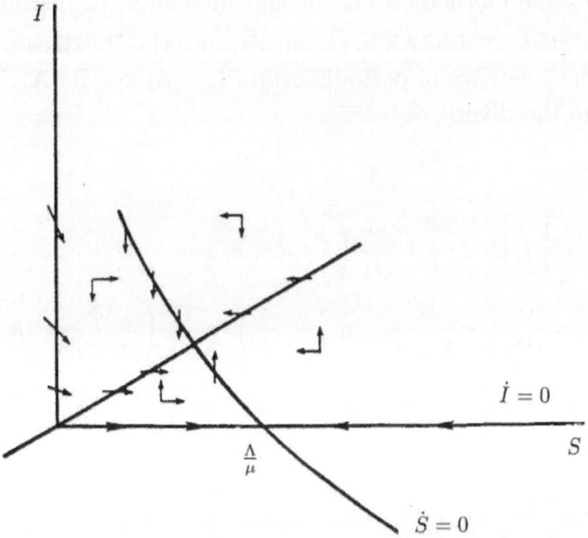

Fig. 3.6. Phase diagram for system (3.50) for $\mu > 0$ and $\lambda - (k+\mu) > 0$ [125]

Theorem 3.8.

a) *If* $\lambda < k + \mu$ *the trivial equilibrium* $z^* := (S^*, I^*, A^*)^T = \left(\frac{\Lambda}{\mu}, 0, 0\right)^T$ *is the only equilibrium solution of system (3.50). It is GAS in* Ω.

b) *If* $\lambda > k + \mu$ *system (3.50) admits two equilibrium points;* $z^* := (S^*, I^*, A^*)^T = \left(\frac{\Lambda}{\mu}, 0, 0\right)^T$ *which is unstable, and* $z^{**} := (S^{**}, I^{**}, A^{**})^T \in \overset{\circ}{\Omega}$ *which is GAS in* $\overset{\circ}{\Omega}$.

Proof. The existence of equilibrium solutions is a consequence of the geometric analysis of the isoclines (3.54), (3.55).

If $\lambda < k+\mu$ GAS of z^* in Ω can be proven by using the Lyapunov function

$$V(S, I) = I \ .$$

If $\lambda > k + \mu$ one can prove directly that the Jacobi matrix of the first two equations in (3.50) at z^{**} has a positive determinant and a negative trace, so that both eigenvalues have negative real part. This means that z^{**} is LAS, while by the same methods it can be shown that z^* is unstable.

To prove that z^{**} is GAS in $\overset{\circ}{\Omega}$, one can use the Bendixson-Dulac criterion to rule out the existence of periodic orbits (see Appendix A, Section A.1.1.2).

Compute the divergence

$$\begin{aligned}
&\frac{\partial}{\partial S}\left(\frac{1}{I} f_1(S, I)\right) + \frac{\partial}{\partial I}\left(\frac{1}{I} f_2(S, I)\right) = \\
&\quad = \frac{\partial}{\partial S}\left(-\frac{\lambda S}{S+I} - \mu\frac{S}{I} + \frac{\Lambda}{I}\right) + \frac{\partial}{\partial I}\left(\frac{\lambda S}{S+I} - (k+\mu)\right) \\
&\quad = -\frac{\lambda}{S+I} - \frac{\mu}{I} < 0 \ , \qquad \text{in} \quad \overset{\circ}{\Omega} \ .
\end{aligned}$$

This completes the proof.

3.4.1.2. Model II [65]

In [65] Castillo-Chavez et al. presented an extension of model (3.50) in which infected individuals may follow two different paths; either they go in a class I, of individuals that will go on to develop AIDS, or they go in a class Y, of individuals that will not develop full blown AIDS.

From class I individuals decay then in class A, of individuals that have developed full blown AIDS ; from class Y individuals decay into class Z, of individuals that even if have not developed full blown AIDS, are no longer sexually active. Note that individuals belonging to either A or Z do not enter any longer into the dynamics of the diseases. We shall denote by $p \in [0,1]$ that fraction of individuals that goes into the I class, so that $1-p$ is the fraction that goes into the Y class (see Fig. 3.7).

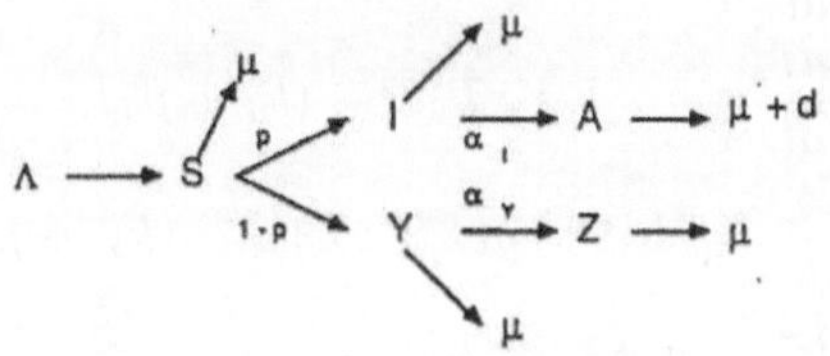

Fig. 3.7. Flow diagram for the model (3.60) [65]

As far as the force of infection is concerned it is assumed that it is given by

$$g(I,Y;T) = \lambda \frac{C(T)}{T}(I+Y) \tag{3.56}$$

where I and Y are the two classes of infectives and

$$T = S + I + Y\ . \tag{3.57}$$

Here the function $C(T)$ denotes the mean number of sexual partners an average individual has per unit time, given the population density is T, and λ (a constant) denotes the average sexual risk per infected partner. The factor $(I+Y)/T$ is the probability that the contact of a susceptible with a randomly selected individual will be with an infectious individual.

The incidence rate of new infectives will be then

$$\textit{total incidence rate} = \lambda \frac{C(T)}{T}(I+Y)S \tag{3.58}$$

that has to be splitted into I by a fraction p and into Y by a fraction $1-p$.

In the sequel it is assumed that the function C is a differentiable and increasing function of $T \in \mathbb{R}_+$:

$$\begin{aligned} C(T) &> 0\,, \qquad \text{for} \quad T \geq 0 \\ C'(T) &\geq 0\,, \qquad \text{for} \quad T > 0 \end{aligned} \tag{3.59}$$

With all this in mind, the epidemic model in [65] is described by the following ODE system

$$\begin{cases} \dfrac{dS}{dt} = \Lambda - g(I,Y;T)\,S - \mu S \\ \dfrac{dI}{dt} = p\,g(I,Y;T)\,S - (k_I + \mu)I \\ \dfrac{dY}{dt} = (1-p)\,g(I,Y;T)\,S - (k_Y + \mu)Y \\ \dfrac{dA}{dt} = k_I I - (\mu + \delta)A \\ \dfrac{dZ}{dt} = k_Y Y - \mu Z \end{cases} \tag{3.60}$$

for $t > 0$.

Again, it is not difficult to show that system (3.60) is well posed when subject to initial conditions $(S^o, I^o, Y^o, A^o, Z^o) \in \mathbb{R}^5_+$.

The total population $N(t) = S(t)+I(t)+Y(t)+A(t)+Z(t)$ is a dynamical variable the evolution equation of which is

$$\frac{dN}{dt} = -\mu N - \delta A + \Lambda\ .$$

It is then possible to show as for Proposition 3.5 that the set

$$\Omega := \left\{ (S,I,Y,A,Z) \in \mathbb{R}^5_+ \;\middle|\; S + I + Y + A + Z \leq \frac{\Lambda}{\mu} \right\} \tag{3.61}$$

is a compact, convex, invariant set for system (3.60).

For $\mu > 0$, system (3.60) admits the disease-free equilibrium

$$S^* = \frac{\Lambda}{\mu}\,, \quad I^* = 0\,, \quad Y^* = 0\,, \quad A^* = 0\,, \quad Z^* = 0\,. \tag{3.62}$$

Observe now that the dynamics of S, I, Y is governed independently of A and Z; therefore it will suffice to analyze the first three equations in (3.60) . If we denote

$$\sigma_I := k_I + \mu \ , \qquad \sigma_Y := k_Y + \mu \tag{3.63}$$

$$D := \left(\frac{p}{\sigma_I} + \frac{1-p}{\sigma_Y} \right) \tag{3.64}$$

$$R := \lambda C \left(\frac{\Lambda}{\mu} \right) D \tag{3.65}$$

the following theorem holds [65].

Theorem 3.9. *Let* $\mu > 0$.

a) If $R \leq 1$ *then* $z^* := \left(\frac{\Lambda}{\mu}, 0, 0, 0, 0 \right)^T$ *is GAS in* Ω.

b) If $R > 1$ *then* z^* *is unstable and there exists a unique nontrivial equilibrium* z^{**} *which is LAS in* $\mathring{\Omega}$.

Corollary 3.10. *Let* $\mu > 0$, *and* $k_I = k_Y$ *or* $p = 1$. *If* $R > 1$ *then* z^{**} *is GAS in* $\mathring{\Omega}$.

For the proofs we directly refer to [65].

3.4.2. One population models. Distributed time of infectiousness

In the previous models (3.50) and (3.60) a linear decay was assumed from the I class into the A class, which corresponds to an exponential survival time in the I class; the decay rate is independent of the length of time that an individual has been infected. Although the distribution of times between infection and the onset of clinical AIDS is only partially known, it appears from available data that the rate of conversion from the I to the A class, or from the Y to the Z class, has a more general distribution (see e.g. [10, 34]).

In order to take this into account Castillo-Chavez et al. in [65] introduce a function $P(s)$ which gives the fraction of infectives remaining infectious a time $s \geq 0$ after becoming infective (the same idea has been explored in the general context of mathematical models for epidemics in [36, 37]).

The function $P : \mathbb{R}_+ \to \mathbb{R}_+$ is a survival function so that we must assume it to be nonincreasing and such that $P(0) = 1$. The quantity

$$\int_0^\infty P(s)\, ds = \tau < \infty \tag{3.66}$$

represents the average sojourn time of an individual in the infective class (if not subject to death). The function $\rho(s) = -\frac{d}{ds}P(s)$ will give the probability density of the sojourn time in the infective class. If we assume $P(s) = e^{-ks}$, $s \geq 0$ we have the classical exponential sojourn time with a linear decay having rate k independent of s. The other extreme is

$$P(s) = \begin{cases} 1 , & 0 \leq s \leq \tau \\ 0 , & \tau < s \end{cases} \tag{3.67}$$

in which case we have the fixed delay $\tau > 0$ for an individual to go from the infective class I to the removed class A. In Sect. 3.4.3 a gamma distribution has been proposed for the incubation period.

With this in mind system (3.60) becomes

$$(3.68) \quad \begin{cases} \dfrac{d}{dt}S(t) = \Lambda - \tilde{g}(t)\, S(t) - \mu S(t) \\ I(t) = I^o(t) + p \displaystyle\int_0^t \tilde{g}(s)\, S(s) e^{-\mu(t-s)} P_I(t-s)\, ds \\ Y(t) = Y^o(t) + (1-p) \displaystyle\int_0^t \tilde{g}(s)\, S(s) e^{-\mu(t-s)} P_Y(t-s)\, ds \\ A(t) = A^o e^{-(\mu+\delta)\,t} + A^o(t) + p \displaystyle\int_0^t \int_0^\tau \tilde{g}(s)\, S(s) e^{-(\mu+\delta)(t-s)} \\ \qquad \times \left[\rho_I(\tau - s) e^{-(\mu+\delta)(t-s)}\right] ds\, d\tau \\ Z(t) = Z^o e^{-\mu t} + Z^o(t) + (1-p) \displaystyle\int_0^t \int_0^\tau \tilde{g}(s)\, S(s) e^{-\mu(t-s)} \\ \qquad \times \left[\rho_Y(\tau - s) e^{-\mu(t-s)}\right] ds\, d\tau \end{cases}$$

where $\tilde{g}(t) := g(I(t), Y(t); T(t))$ is the force of infection as a function of time, for $t \geq 0$; $I^o(t)$ and $Y^o(t)$ denote those individuals that were in either class I or Y at time $t = 0$, and are still infectious (they are assumed to vanish for large enough $t > 0$); $A^o = A(0)$ represents those who had full blown AIDS at time zero and are still alive; $A^o(t)$ represents those initially in class I who have moved in class A and are still alive at time t (it is assumed that $A^o(t)$ approaches zero as t approaches infinity); Z^o and $Z^o(t)$ are the corresponding term for the other group of infectives. In (3.68) the factors $e^{-\mu(t-s)}$ and $e^{-(\mu+\delta)\,(t-s)}$ take into account removals due to death.

System (3.68) is a system of nonlinear integro-differential equations. Its well posedness has to be proved by specific techniques which are beyond the interests of this presentation (see [65, 36, 37]).

We shall report the results obtained in [65] without proofs.

Let in this case denote

$$(3.69) \qquad D := \int_0^\infty \left[p\, P_I(s) + (1-p) P_Y(s)\right] e^{-\mu s}\, ds$$

$$(3.70) \qquad R := \lambda C \left(\frac{\Lambda}{\mu}\right) D\,.$$

The dynamics of the classes S, I, Y is governed autonomously, so that we can restrict as usual our analysis to the first three equations of system (3.68).

If $I^o(t) = Y^o(t) = 0$ the reduced system always has the trivial disease free equilibrium $z^* := (S^*, I^*, Y^*) = \left(\dfrac{\Lambda}{\mu}, 0, 0\right)$; otherwise it does not have a constant solution. But since we have assumed that $I^o(t)$ and $Y^o(t)$ are zero for large t , the following theorem holds [65]

Theorem 3.11.

a) *If $R \leq 1$ then the trivial state z^* is a global attractor for system (3.68); i.e. for any positive solution of system (3.68) we have*

$$\lim_{t \to +\infty} (S(t), I(t), Y(t)) = \left(\frac{\Lambda}{\mu}, 0, 0 \right) .$$

b) *If $R > 1$, then z^* is unstable. Furthermore there exists a constant $W^{**} > 0$ such that any positive solution of system (3.68) satisfies*

$$\limsup_{t \to +\infty} [I(t) + Y(t)] \geq W^{**}$$

.

Under the assumption that $I^o(t)$ and $Y^o(t)$ are zero for large t the authors [65] conjecture that $S(t), I(t), Y(t)$ satisfy for large t the limiting system

$$\text{(3.71)} \quad \begin{cases} \dfrac{dS}{dt} = \Lambda - g(I, Y; T)\, S - \mu S \\ I(t) = p \displaystyle\int_{-\infty}^{t} g(I(s), Y(s); T(s))\, S(s) e^{-\mu(t-s)} P_I(t-s)\, ds \\ Y(t) = 1 - p \displaystyle\int_{-\infty}^{t} g(I(s), Y(s); T(s))\, S(s) e^{-\mu(t-s)} P_Y(t-s)\, ds \end{cases}$$

so that the quantity

$$W(t) := I(t) + Y(t) , \qquad t \geq 0$$

satisfies, for large t,

$$\text{(3.72)} \qquad W(t) = \int_{-\infty}^{t} g(I(s), Y(s); T(s))\, S(s)\, e^{-\mu(t-s)} P(t-s)\, ds$$

where

$$\text{(3.73)} \qquad P(s) = p\, P_I(s) + (1-p)\, P_Y(s) , \qquad s \geq 0$$

For system (3.71), (3.72) the following theorem has been proven [65].

Theorem 3.12. *If $R > 1$, then system (3.71), (3.72) has a unique nontrivial equilibrium $(S^{**}, W^{**}) \gg 0$. If in addition $\frac{d}{dT}\frac{C(T)}{T} \leq 0$, then this endemic equilibrium is LAS.*

3.4.3. One population models. Multiple stages of infection, with variable infectiousness

We wish to point out that in the models introduced above it has been assumed not only homogeneous mixing, but also that an infectious individual has always the same infectiousness, independently of the time at which he has got the infection.

Actually this assumption is not realistic, especially if we assume a distributed time of infectiousness (for a discussion about variable infectiousness in epidemic models with respect to the disease-age we refer to [47, 132;Part I, 224]; for medical evidence we refer to [88, 144, 194]).

In this paragraph we give an account of a model with gamma distributed time of infectiousness and differentiated infectiousness , as proposed in [66]. A possible choice of the distribution of the residence time in the infectious class I is the gamma distribution (see [157] for an epidemiological discussion) whose density is given by

$$(3.74) \qquad \rho(s) = \frac{k^m\, t^{m-1}\, e^{-ks}}{(m-1)!} \,, \qquad s \geq 0 \,.$$

As also shown in Sect. 2.4.2 [30] this case can be handled equivalently (as far as the asymptotic behavior is concerned) by considering m stages of infectiousness with exponential decay with the same rate k among the different stages. Different decay rates would imply general multiparameter gamma distributions, as those identified by Longini et al. [157].

This case has been considered in [125] as an extension of model (3.50). On the other hand the introduction of multiple stages makes it possible to consider different forces of infection for the different stages so as to overcome the limitations of the previous models.

Starting from model (3.50) the I class is splitted into $m \geq 1$ subclasses I_j, $j = 1, \ldots, m$. The force of infection due to the j-th class of infectives is given by

$$(3.75) \qquad g_j(I_1, \ldots, I_m; T) = \frac{\lambda_j}{T} I_j \,, \qquad j = 1, \ldots, m$$

where now

$$T := I + S := \sum_{j=1}^{m} I_j + S \, , \tag{3.76}$$

so that the total force of infection acting on each susceptible individual is

$$g(I_1, \ldots, I_m; T) = \sum_{j=1}^{m} g_j(I_1, \ldots, I_m; T) = \sum_{j=1}^{m} \frac{\lambda_j}{T} I_j \tag{3.77}$$

The epidemic model is thus given by

$$\begin{cases} \dfrac{dS}{dt} = -g(I_1, \ldots, I_m; T)\, S - \mu S + \Lambda \\ \dfrac{dI_1}{dt} = g(I_1, \ldots, I_m; T)\, S - (k + \mu) I_1 \\ \dfrac{dI_j}{dt} = k I_{j-1} - (k + \mu) I_j \, , \qquad j = 2, \ldots, m \\ \dfrac{dA}{dt} = k I_m - (\mu + \delta) A \end{cases} \tag{3.78}$$

As usual we limit our analysis to the first $m + 1$ equations in system (3.78).

The trivial equilibrium (no-disease equilibrium) of such system is given by

$$z^* := (S^*, I_1^*, \ldots, I_m^*)^T := \left(\frac{\Lambda}{\mu}, 0, \ldots, 0\right)^T . \tag{3.79}$$

We make a shift of coordinates

$$X = S - S^* = S - \frac{\Lambda}{\mu},$$

so that the evolution system for $x := (X, I_1, \ldots, I_m)^T$ is

$$\frac{dx}{dt} = \mathcal{A}x + Q(x) \tag{3.80}$$

where

$$\mathcal{A} = \begin{pmatrix} -\mu & \lambda_1 & \lambda_2 & \dots & \lambda_{m-1} & \lambda_m \\ 0 & \lambda_1 - (k+\mu) & \lambda_2 & \dots & \lambda_{m-1} & \lambda_m \\ 0 & k & -(k+\mu) & \dots & 0 & 0 \\ \vdots & \vdots & \vdots & \ddots & \vdots & \vdots \\ 0 & 0 & 0 & \dots & k & -(k+\mu) \end{pmatrix} \tag{3.81}$$

$$=: \left(\begin{array}{c|c} -\mu & * \\ \hline 0 & \mathcal{A}_1 \end{array} \right)$$

and

$$Q(x) = \left(-\frac{I}{T} \sum_{j=1}^{m} \lambda_j I_j,\ -\frac{I}{T} \sum_{j=1}^{m} \lambda_j I_j,\ 0, \dots, 0 \right)^T \tag{3.82}$$

The following properties hold for the matrix $\mathcal{A}_1$ and for the vector function Q.

i) $\mathcal{A}_1$ is an irreducible compartmental matrix, in that all off-diagonal entries are nonnegative.

ii) $Q(x) = o(x)$, i.e. $\lim\limits_{x \to 0} \dfrac{Q(x)}{\|x\|} = 0$

iii) $Q(x) \leq 0$ for all x in the domain of definition.

As a consequence of i), the Perron-Frobenius theorem (see Appendix A, Section A.4.1) implies that $\mathcal{A}_1^T$ has a dominant real eigenvalue r (with largest real part), to which it corresponds a unique eigenvector $v \gg 0$ in $\mathbb{R}^m$. As a consequence the following theorem holds.

Theorem 3.13. [125]

a) When $r < 0$ the trivial equilibrium $z^* := (\Lambda/\mu, 0, \dots, 0)^T$ is GAS.

b) When $r > 0$, z^* is unstable and a unique nontrivial endemic state $z^{**} \gg 0$ exists which is LAS.

Proof. For part a), if $r < 0$, then $\mathcal{A}_1^T$, $\mathcal{A}_1$ and $\mathcal{A}$ have only eigenvalues with negative real part, so that the point $x^* := (0, 0, \ldots, 0)^T$ is LAS.

For the GAS, one can make use of the Lyapunov function

$$V(X, I_1, \ldots, I_m) := v^T (I_1, \ldots, I_m)^T$$

so that its derivative along the trajectories of system (3.80) is

$$\begin{aligned}\dot{V}(X, I_1, \ldots, I_m) &= v^T \mathcal{A}_1 (I_1, \ldots, I_m)^T + (0, v^T) \cdot Q(X, I_1, \ldots, I_m) \\ &= r v^T (I_1, \ldots, I_m)^T + (0, v^T) \cdot Q(X, I_1, \ldots, I_m) \leq 0\end{aligned}$$

Furthermore the set of points at which $\dot{V} = 0$ is the subspace

$$\{(I_1, \ldots, I_m)^T \in \mathbb{R}^m \mid I_j = 0, \quad j = 1, \ldots, m\}$$

on which system (3.80) tends globally to $(X^*, I_1^*, \ldots, I_m^*) = (0, 0, \ldots, 0)$.

For part b), when $r > 0$ the origin is unstable. It can be shown by direct calculation that in this case a unique endemic solution z^{**} exists for system (3.80), and further that it is LAS by the analysis of the Jacobi matrix of system (3.78) at z^{**} [125].

An explicit expression of the "threshold parameter" in terms of the biological parameters of system (3.78) is given in Sect. 3.4.4 (see Table 3.7).

3.4.4. Multigroup models with multiple stages of infectiousness

In the mathematical modelling of AIDS it is of great relevance to divide the total population at risk into n groups to take into account the pattern of contacts between different groups that differ in sexual activity.

The population at risk is in fact heterogeneous in a variety of ways. The contacts between people can be homosexual, heterosexual, or by needle sharing among intravenous drug users; some groups have higher contact rates than others, etc.

Models that take into account all these facts are now available by different authors (see the bibliography at the end of the volume). Here we shall discuss a general form of the force of infection which includes most of the possible patterns of contacts, and later we shall present some results selected from the literature.

Clearly we cannot report about all existing literature on the subject, so that the presentation will reflect the "leit motiv" of the present monograph.

Suppose we have n groups and that in the i-th group $(i = 1, \ldots, n)$ we identify a class S_i of susceptibles, m stages of infectiousness I_{ir}, $r = 1, \ldots, m$, and a class A_i of individuals who have fully developed AIDS and so do not participate any more to the dynamics of the disease. The force of infection acting on each of the S_i susceptibles will be given in general by

$$(3.83) \qquad \begin{aligned} g_i &:= g_i\left(I_{jr}\,,\; j = 1, \ldots, n;\, r = 1, \ldots, m;\, T_j,\; j = 1, \ldots, n\right) \\ &:= c_i \sum_{j=1}^{n} \rho_{ij} \sum_{r=1}^{m} \beta_{ijr} \frac{I_{jr}}{T_j}\,, \qquad i = 1, \ldots, n\,, \end{aligned}$$

where

$$(3.84) \qquad T_j = S_j + I_j\,, \qquad j = 1, \ldots, n\,,$$

with

$$(3.85) \qquad I_j = \sum_{r=1}^{m} I_{jr}\,, \qquad j = 1, \ldots, n\,.$$

Note that T_j denotes the whole sexually active population in the j-th group $(j = 1, \ldots, n)$.

In (3.83) $c_i = c_i(T)$ is the mean number of sexual partners that an average individual has per unit time. For very small populations it may be a function of the total sexually active population

$$T = \sum_{j=1}^{n} T_j \ . \tag{3.86}$$

Usually it is assumed to be a constant [122, 125].

The structure of the matrix $P := (\rho_{ij})_{i,j=1,\dots,n}$ defines different patterns of contacts between groups. Three main cases have been identified; we have

Case I : Restricted mixing [125], when all contacts are restricted to within group contacts;

$$\rho_{ii} = 1\,, \quad i = 1, \dots, n\,; \qquad P = I_{n \times n} \tag{3.87}$$

In this case each subpopulation can be analyzed independently of each other (see Sect. 3.4.3).

Case II : Proportionate (or random) mixing [176, 115], when the fraction of contacts of group i with group j is equal to the fraction of total contacts made by the population that are due to group j $(i, j = 1, \dots, n)$. Now, since the total number of contacts by all groups is

$$C(T_1, \dots, T_k) = \sum_{k=1}^{n} c_k(T)\, T_k \ , \tag{3.88}$$

we have in this case

$$\rho_{ij} = c_j(T) \frac{T_j}{C(T_1, \dots, T_k)}\ , \qquad i, j = 1, \dots, n\ . \tag{3.89}$$

All the rows in the matrix P are the same.

Case III : Preferred (or biased) mixing [115, 166, 165, 118, 122, 125, 66], when part of the contacts are reserved to within group contacts; the nonreserved contacts are subject to proportionate mixing. The matrix P has the following entries

$$\rho_{ii} = \rho_i + (1 - \rho_i) \frac{c_i(T)\,(1 - \rho_i)\,T_i}{\sum_{k=1}^{n} c_k(T)\,(1 - \rho_k)\,T_k} \tag{3.90}$$

$$\rho_{ij} = (1-\rho_j)\,\frac{c_j(T)\,(1-\rho_j)\,T_j}{\sum_{k=1}^{n} c_k(T)\,(1-\rho_k)\,T_k}\,, \qquad i \neq j \tag{3.91}$$

for $i, j = 1, \ldots, n$. The quantity ρ_i, $i = 1, \ldots, n$, is the fraction of group i's contacts reserved for within group contacts. Clearly

$$\begin{aligned} \rho_i = 0\,, &\quad \text{for all } i \iff \text{proportionate mixing} \\ \rho_i = 1\,, &\quad \text{for all } i \iff \text{restricted mixing}\,. \end{aligned}$$

The entries of the matrix $P = (\rho_{ij})_{i,j=1,\ldots,n}$ must clearly satisfy a symmetry constraint ; the number of contacts per unit time of group i with group j must equal the number of contacts of j with i, giving

$$c_i(T)\,T_i\,\rho_{ij}(T_1,\ldots,T_n) = c_j(T)\,T_j\,\rho_{ij}(T_1,\ldots,T_n)\,, \quad i,j = 1,\ldots,n\,. \tag{3.92}$$

Finally the last quantity in (3.83)

$$g_{ij}(I_{j1},\ldots,I_{jm};T_j) := \sum_{r=1}^{m} \beta_{ijr}\,\frac{I_{jr}}{T_j}\,, \qquad i,j = 1,\ldots,n\,, \tag{3.93}$$

is the contribution to the force of infection acting on the susceptibles S_i of the i-th group ($i = 1, \ldots, n$), due to the infectives $I_{j1}, \ldots, I_{jm}$ of the j-th group ($j = 1, \ldots, n$).

It corresponds to the quantity defined in (3.77) for within group infection.

Hence β_{ijr} plays the role of the transmission coefficient between the infectives I_{jr} and the susceptibles S_i ($i, j = 1, \ldots, n$; $r = 1, \ldots, m$).

If one takes (3.83) into account as the force of infection acting on the susceptibles of the i-th group ($i = 1, \ldots, n$), model (3.78) can be extended to the multigroup case as follows

$$\begin{cases} \dfrac{d}{dt}S_i = -g_i S_i - \mu S_i + \Lambda_i \\ \dfrac{d}{dt}I_{i1} = g_i S_i - (k_i+\mu)\,I_{i1} \\ \dfrac{d}{dt}I_{ir} = k_i I_{i,r-1} - (k_i+\mu)\,I_{ir}\,, \qquad r = 2,\ldots,m \\ \dfrac{d}{dt}A_i = k_i I_{im} - (\delta_i+\mu)A_i \end{cases} \tag{3.94}$$

for $i = 1, \ldots, n$ and $t > 0$.

If we sum up the equations in system (3.94) for any $i = 1, \ldots, n$, we get the evolution equation for the total population in the i-th group.

$$\frac{d}{dt} N_i = -\mu N_i + \Lambda_i - \delta_i A_i \,, \qquad i = 1, \ldots, n \ . \tag{3.95}$$

As a direct consequence, system (3.94) satisfies the following properties [125] which directly correspond to the results in Propositions 3.5, 3.6, 3.7 in Sect. 3.4.1 .

Proposition 3.14. *For $\mu > 0$, let*

$$N_i(t) := S_i(t) + I_i(t) + A_i(t) \,, \qquad t \geq 0, \quad i = 1, \ldots, n.$$

If $N_i(0) \leq \dfrac{\Lambda_i}{\mu}$,then $N_i(t) \leq \dfrac{\Lambda_i}{\mu}$, for all $t \geq 0$. Furthermore the set

$$\Omega := \left\{ (S_1, \ldots, A_n) \in \mathbb{R}_+^{n+n\cdot m+n} \mid S_i + I_i + A_i \leq \frac{\Lambda_i}{\mu}, i = 1, \ldots, n \right\} \tag{3.96}$$

is a compact, convex, invariant set for system (3.94).

Proposition 3.15. *$N_i(t)$ is, for any $i = 1, \ldots, n$, a dynamical variable. If $\delta_i = 0$, $i = 1, \ldots, n$ in system (3.94), then $N_i(t)$ satisfies*

$$N_i(t) = \left(N_i(0) - \frac{\Lambda_i}{\mu} \right) e^{-\mu t} + \frac{\Lambda_i}{\mu} \,, \qquad i = 1, \ldots, n \,, \quad t \geq 0 \,, \tag{3.97}$$

and

$$\lim_{t \to +\infty} N_i(t) = \frac{\Lambda_i}{\mu} \,, \qquad i = 1, \ldots, n \ . \tag{3.98}$$

Proposition 3.16. *For $\mu > 0$, system (3.94) admits the disease-free equilibrium*

$$S_i^* = \frac{\Lambda_i}{\mu} \,, \quad I_{i1}^* = \cdots = I_{im}^* = A_i^* = 0 \,, \qquad i = 1, \ldots, n. \tag{3.99}$$

About the existence of nontrivial endemic states and the stability of the equilibria we report here the results of Castillo-Chavez et al. [66], for the case of only one stage of infection ($m = 1$).

Let,

$$\theta_i(T) := \rho_i \beta_{ii} c_i(T) \tag{3.100}$$

$$r_i(T) := (1 - \rho_i)\, c_i(T) \tag{3.101}$$

$$\begin{aligned} l_{ij}(T) &:= c_i(T)\,(1-\rho_i)\, c_j(T)\,(1-\rho_j)\,\beta_{ij} \\ &= r_i(T)\, r_j(T)\, \beta_{ij} \end{aligned} \tag{3.102}$$

for $i, j = 1, \ldots, n$, and introduce the matrices $Q(\mu)$ and $H(\mu)$ given by

$$Q(\mu) := diag\left(\frac{\theta_i(T^*)}{\sigma_i + 1}\right) + diag\left(\frac{\Lambda_i}{K(T^*)\,(\sigma_i+1)}\right) L(T^*) \tag{3.103}$$

$$H(\mu) := Q(\mu) - \mu I_{n \times n} \tag{3.104}$$

where

$$T^* = \frac{1}{\mu}\sum_{k=1}^{n} \Lambda_k\,; \qquad K(T^*) = \sum_{k=1}^{n} r_k(T^*)\,\Lambda_k\,;$$

$$\sigma_i = \frac{k_i}{\mu} \quad i = 1, \ldots, n; \qquad L(T^*) = (l_{ij}(T^*))_{i,j=1,\ldots,n}$$

and $I_{n\times n}$ is the $n \times n$ identity matrix. Now let , for any $\mu > 0$,

$$M(H(\mu)) := \sup\left\{ Re\ \rho \mid \det\left(\rho I_{n\times n} - H(\mu)\right) = 0 \right\}. \tag{3.105}$$

The following lemma holds [66, 67].

Lemma 3.17. *Under the above assumptions, if furthermore*

$$\frac{d}{dT}c_i(T) \geq 0\,; \quad \frac{d}{dT}\left(\frac{c_i(T)}{T}\right) \leq 0\,, \qquad i = 1, \ldots, n \tag{3.106}$$

for $T > 0$, then there exists a unique $\mu_o > 0$ such that

$$M(H(\mu)) \begin{cases} < 0 & \text{if} \quad \mu > \mu_o \\ = 0 & \text{if} \quad \mu = \mu_o \\ > 0 & \text{if} \quad \mu < \mu_o\,. \end{cases} \tag{3.107}$$

Further the sign of $M(H(\mu))$ implies the stability properties of the trivial steady state (3.99).

Proposition 3.18. *Under the same assumptions of Lemma 3.17, the infection free state (3.99) is LAS if $M(H(\mu)) < 0$; it is unstable if $M(H(\mu)) > 0$.*

By the methods of bifurcation analysis, Castillo-Chavez et al. [67, 122] have also given information about the existence and stability of nontrivial endemic equilibria, under the additional assumptions that $c_i(T) = c_i$ (constants) for $i = 1, \ldots, n$, and $Q(\mu)$ irreducible.

If μ_o is such that $M(H(\mu_o)) = 0$ (see (3.107)), we introduce the expression

$$h(\mu_o) = \sum_{i=1}^{n} \overline{I}_i I_i \Bigg[\frac{K I_i(\mu_o(\sigma_i + 1)^2 - \theta_i \sigma_i}{\Lambda_i^2} - \left(\sum_{j=1}^{n} r_j \sigma_j I_j\right) \frac{\mu_o(\sigma_i + 1) - \theta_i}{\Lambda_i}\Bigg] \tag{3.108}$$

where $I = (I_1, \ldots, I_n)^T$, respectively $\overline{I} = (\overline{I}_1, \ldots, \overline{I}_n)^T$, is the positive eigenvector of $H(\mu_o)$, respectively of $[H(\mu_o)]^T$ corresponding to their zero eigenvalue. The existence of these strictly positive eigenvectors is a consequence of the Perron-Frobenius theory for irreducible M-matrices (see Appendix A, Section A.4.1 and [31]).

The following main theorem holds [66, 67, 122].

Theorem 3.19. *Suppose $c_i(T) = c_i$ (constants), for $i = 1, \ldots, n$; Q is irreducible, and μ_o is such that $M(H(\mu_o)) = 0$. Let $h(\mu_o)$ be defined as in (3.108).*

a) *For each* $\mu \in (0, \mu_o)$, *system (3.94) admits at least one positive endemic equilibrium.*

b) *If* $h(\mu_o) < 0$ *then there is an* $\epsilon > 0$ *such that, for any* $\mu \in (\mu_o, \mu_o + \epsilon)$, *system (3.94) admits at least two positive endemic equilibria, one of which is unstable.*

c) *If* $h(\mu_o) > 0$ *then there is an* $\epsilon > 0$ *such that for any* $\mu \in (\mu_o - \epsilon, \mu_o)$ *system (3.94) admits at least one positive equilibrium which is LAS.*

A refinement of Theorem 3.19 a) has been obtained by Lin [154] for the multigroup model (3.94), but only one stage of infectiousness.

Specifically Lin considers the following model

$$(3.109) \qquad \begin{cases} \frac{d}{dt}S_i = -g_i S_i - \mu S_i + \Lambda_i \\ \frac{d}{dt}I_i = g_i S_i - (k_i + \mu)I_i \\ \frac{d}{dt}A_i = k_i I_i - (\delta_i + \mu)A_i \end{cases}$$

for $i = 1, \cdots, n$ and $t > 0$, under the assumption of proportionate mixing. This means that

$$g_i := c_i \sum_{j=1}^{n} \rho_{ij}\beta_{ij}\frac{I_j}{T_j} \quad , \qquad i = 1, \cdots, n \ ,$$

with ρ_{ij} given by (3.89) where it is further assumed as in Theorem 3.19 that the c_i's $(i = 1, \cdots, n)$ are constants.

The following system is introduced

$$(3.110a) \qquad \frac{d}{dt}v_i = -\mu v_i + k_i I_i$$

$$(3.110b) \qquad \frac{d}{dt}I_i = -(k_i + \mu)I_i + c_i(\Lambda_i - \mu v_i - \mu I_i)\frac{\sum_{j=1}^{n} c_j\beta_{ij}I_j}{\sum_{k=1}^{n} c_k(\Lambda_k - \mu v_k)}$$

for $i = 1, \cdots, n, \quad t > 0.$

System (3.110) is equivalent to system (3.109) provided $v_i \geq 0$, $I_i \geq 0$, $v_i + I_i \leq \frac{\Lambda_i}{\mu}$ $(i = 1, \cdots, n)$, via the relations

$$v_i = \frac{\Lambda_i}{\mu} - S_i - I_i \ , \qquad i = 1, \cdots, n \, .$$

Let

$$K := diag\{k_1, \cdots, k_n\}$$

$$L := \frac{1}{\sum_{k=1}^{n} c_k \Lambda_k} \begin{pmatrix} c_1^2 \Lambda_1 \beta_{11} & \cdots & c_1 \Lambda_1 c_n \beta_{1n} \\ \vdots & \ddots & \vdots \\ c_n \Lambda_n c_1 \beta_{n1} & \cdots & c_n^2 \Lambda_n \beta_{nn} \end{pmatrix} ,$$

and

$$A(\mu) := -K - \mu E + L$$

where $E = I_{n \times n}$, the identity matrix.

The Jacobian of the right hand side of (3.110) at the disease-free equilibrium (in this case the origin) is

$$J(\mu) := \begin{pmatrix} -\mu E & K \\ 0 & A(\mu) \end{pmatrix}$$

so that the stability of the disease-free equilibrium is decided by the stability properties of $A(\mu)$.

From now on we assume, as in Theorem 3.19, that L is irreducible so that also $A(\mu)$ is such.

The bifurcation parameter μ is such that

$$s(A(\mu)) \begin{cases} > 0 & \text{for} \ \ \mu \in (0, \mu_0) \\ = 0 & \text{for} \ \ \mu = \mu_0 \\ < 0 & \text{for} \ \ \mu > \mu_0 \ , \end{cases}$$

where $s(A(\mu))$ denotes the stability modulus of $A(\mu)$, i.e.

$$s(A(\mu)) := sup\{Re\, \lambda \mid \lambda \text{ eigenvalue of } A(\mu)\} \, .$$

The following result holds true [66, 122, 153]

Theorem 3.20. *Suppose $c_i(T) = c_i$ (constants), for $i = 1, \cdots, n$; L is irreducible, and μ_0 is such that $s(A(\mu_0)) = 0$. If $A(\mu)$ is a nonnegative matrix, then the system (3.110) admits a unique positive equilibrium for any $\mu \in (0, \mu_0)$.*

We refer to [154] for a detailed proof of the previous theorem, since it requires various preparatory lemmas which take into account the cooperative structure of system (3.110) (see Chapter 4).

It is interesting to remark that the condition on $A(\mu)$ being nonnegative is equivalent to

$$R_i \geq 1 \ , \qquad i = 1, \cdots, n \ . \tag{3.111}$$

where

$$R_i := \frac{\hat{c}_i \beta_{ii} c_i}{\mu + k_i} \ , \qquad i = 1, \cdots, n \ ,$$

with

$$\hat{c}_i := \frac{c_i \Lambda_i}{\sum_{k=1}^{n} c_k \Lambda_k} \ , \qquad i = 1, \cdots, n \ .$$

Hence, epidemiologically, condition (3.111) means that any infectious individual produces at least another infectious individual in his/her own subpopulation over the average incubation period.

Note that also Theorem 3.19 c) has been improved by Lin [154] to ensure again the uniqueness of the positive equilibrium.

The method can be extended (see [154]) to the case of preferred mixing (3.90), (3.91), in which case conditions (3.111) become

$$\frac{c_i^2 (1 - \rho_i)^2 \Lambda_i \beta_{ii}}{(\mu + k_i) \sum_{k=1}^{n} c_k (1 - \rho_k) \Lambda_k} \geq 1 \ , \qquad i = 1, \cdots, n \ . \tag{3.112}$$

Similarly to the case of proportionate mixing, the left hand side of (3.112) is the basic reproduction number within the fraction of the i-th subpopulation's new sexual partners who are distributed according to proportionate mixing.

Further extensions to the case of multigroup models with distributed incubation and variable infectiousness have been analyzed in [184].

It is an easy conjecture that when system (3.94) admits two nontrivial endemic equilibria in addition to the trivial one, the stability pattern should be : the trivial solution LAS; one of the two endemic states LAS; the other one unstable, so that a saddle point behavior may appear. This is confirmed by Theorem 2.3 in [154].

Theorem 3.19 does not relate the existence of nontrivial endemic state to the biological parameters of system (3.94). It is then worth reporting about the results obtained by Jacquez et al. [125], about the "threshold parameter", for such an existence.

They refer to the multistage case ($m \geq 1$), and make the following simplifying assumptions

(i) $\beta_{ijr} = \beta_r$, depending only on the stage of infection and not upon groups $(i, j = 1, \ldots, n\,,\, r = 1, \ldots, m)$;

(ii) $c_i(T) = c_i$, constants, for $i = 1, \ldots, n$;

(iii) $k_i = k$, independent of the group $(i = 1, \ldots, n)$.

For $\mu > 0$, we can define

$$\alpha := \frac{k}{k+\mu}$$

$$\theta_\mu := \frac{1-\alpha^m}{\mu}$$

$$\overline{\beta}_\mu := \sum_{r=1}^{m} \frac{\dfrac{k^{r-1}}{(k+\mu)^r}}{\dfrac{1}{\mu}\left[1-\left(\dfrac{k}{k+\mu}\right)^m\right]}\,\beta_r = \sum_{r=1}^{m} \frac{k\alpha^r}{\theta_\mu}\,\beta_r$$

After long and tedious calculation, they prove the following "threshold theorem" [125]

Theorem 3.21. *Under the assumptions (i)-(iii) listed above, for $\mu > 0$, there is a "threshold parameter" F , depending upon the parameters of system (3.4.48) such that*

i) if $F < 1$ the no-disease equilibrium (3.99) is the only steady state of system (3.94);

ii) if $F > 1$ in addition to the no-disease equilibrium (3.99) a unique nontrivial endemic equilibrium exists for system (3.94) in the interior of Ω

iii) for the case of restricted mixing, when $F < 1$ the no-disease equilibrium (3.99) is GAS, and when $F > 1$ the no-disease equilibrium becomes unstable and the unique endemic equilibrium is LAS.

The expression of F for the three cases of restricted mixing, proportionate mixing, preferred mixing are given in Table 3.7 below.

Proof. For the proofs of i), ii), we refer to [125]; while iii) is a direct consequence of Theorem 3.13 (further analysis can be found in [153]).

Remark . We would like to emphasize that the results of Theorem 3.21 are not contradictory with those of Theorem 3.19 . Take into account the fact that in Theorem 3.21 all the β_{ijr} have been considered to be independent of both i and j .

Table 3.7. Threshold parameter F for $\mu > 0$

Mixing	Threshold parameter
Restricted	$F_i = \overline{\beta}_\mu \theta_\mu c_i$ for group $i = 1, \ldots, n$
Proportionate	$F = \overline{\beta}_\mu \theta_\mu \dfrac{\sum_{i=1}^n c_i^2 \Lambda_i}{\sum_{i=1}^n c_i \Lambda_i}$
Preferred	$F = \overline{\beta}_\mu \theta_\mu \dfrac{\sum_{i=1}^n c_i^2 (1-\rho_i)^2 \Lambda_i (\max\{0, 1 - c_i \overline{\beta}_\mu \theta_\mu \rho_i\})^{-1}}{\sum_{i=1}^n c_i (1-\rho_i) \Lambda_i}$

We may conclude this section by remarking that once again the structure of the basic SIR type model (3.50) with vital dynamics is rather robust concerning the pattern of the equilibria and their stability properties, with respect to modifications of the force of infection that take into account the dependence upon the population of sexually active individuals, with different patterns of contacts; but also with respect to the dependence of the force of infection upon different stages of the disease.

We wish to emphasize that this is a very important property of a mathematical model, since we are still obliged to introduce simplifications with respect to reality, in order to deal with manageable systems but also because of simple ignorance of particular phenomena.

The robustness of the model is later of great importance when dealing with problems of parameter identification.

4. Quasimonotone systems. Positive feedback systems. Cooperative systems

4.1 Introduction

In the previous chapters we have been trying to identify the general structure of epidemic systems and we discovered that they fall essentially in two classes depending on the structure of the interacting matrix in bilinear forces of infection.

The key tool in the analysis was the Lyapunov function.

Actually one of the two classes (class B) can be revisited in another context if we emphasize the fact that cooperative systems, as the venereal, MEM (man- environment-man),... systems are, may be classified with respect to another structure: the one of quasimonotone systems or, using a system theoretic form, positive feedback systems.

The basic property of these systems consists of the fact that comparison theorems apply, or in other words they preserve order.

The mathematical literature has nowadays quite a long tradition, even if it was exploited mostly in the context of parabolic systems [181]. Some authors may like to refer to Kamke [14], but a quite complete analysis is due to Krasnoselskii and his coworkers, for ODE's but also for PDE's [139, 134, 135, 138].

More recently a nice presentation of the theory of quasimonotone parabolic systems is due to Sattinger [196], Martin [161] and Selgrade [199].

A unified approach of the theory of monotone flows is due to M. Hirsch [119] (see also [203]).

Applications of this theory to epidemic systems can be found in [142, 18], among others. In [57, 59, 46, 55] large attention has been devoted to positive-feedback epidemic systems also checking the robustness of the basic model with respect to modifications which take into account realistic features of the system: space structure, periodicity of the parameters, boundary feedback, etc.

This kind of robustness is of great interest for a modeler in that it may imply a structural stability of the model with respect to features that have not been analyzed yet or else that cannot be identified for lack of available data.

We shall start our analysis by listing a series of epidemic systems that may fall in this class. In Appendices A and B, Sections A.4 and B.2.1 respectively, we shall include the main mathematical theorems, that allow the analysis of positive feedback epidemic systems.

4.2. The spatially homogeneous case

In the space and time homogeneous case the system can be described by a set at autonomous ODE equations:

$$\frac{dz}{dt} = f(z) \; , \qquad t > 0 \tag{4.1}$$

where

$$z := (z_1, \ldots, z_m)^T \, , \quad m \in \mathbb{N} - \{0\} \quad \text{and} \quad f : \mathbb{R}^m \to \mathbb{R}^m .$$

We shall suppose in the sequel that f is such that the initial value problem associated with (4.1) is well posed for any choice of the initial condition

$$z^o \in \mathbb{R}^m_+ := \{ z \in \mathbb{R}^m \mid z_i \geq 0, \quad i = 1, \ldots, m \} \, ;$$

i.e. a unique solution $z(t)$ exists for any time $t \geq 0$.

It makes sense then to consider the evolution operator

$$V(t) : \mathbb{R}^m_+ \to \mathbb{R}^m, \quad t \in \mathbb{R}_+, \qquad \text{such that}$$

$$\text{for any} \quad t \in \mathbb{R}_+ : \quad V(t) z^o := z(t)$$

is the solution of (4.1) subject to the initial condition

$$z(0) = z^o \in \mathbb{R}^m_+ \tag{4.2}$$

An equilibrium point for system (4.1) will be any $z^* \in \mathbb{R}^m$ such that

$$\text{for any} \quad t \in \mathbb{R}_+ : \quad V(t) z^* = z^* \tag{4.3}$$

i.e. a fixed point for the whole family of operators

$$\{ V(t), \quad t \in \mathbb{R}_+ \} \; .$$

We say that f is quasimonotone nondecreasing with respect to the cone $\mathbb{K} := \mathbb{R}_+^m$ iff (for the notation refer to Appendix A, Section A.4)

$$\text{for any} \quad j = 1, \ldots, m; \ \text{for any} \quad \xi, \eta \in \mathbb{K}, \ \ \xi \geq \eta : \tag{4.4}$$

$$\xi_j = \eta_j \quad \Longrightarrow \quad f_j(\xi) \geq f_j(\eta)$$

In other words for any $j = 1, \ldots, m,$ $\quad f_j(\xi)$ is monotonically nondecreasing with respect to $\xi_i, \quad i \neq j$.

If we assume that f is continuously differentiable in the interior of $\mathbb{K}$, then an equivalent condition for (4.4) will be

$$\text{for any} \quad i, j, = 1, \ldots, m; \quad i \neq j : \quad \frac{\partial f_i}{\partial \xi_j} \geq 0 \tag{4.5}$$

i.e. the Jacobi matrix $J_f(\xi)$ has nonnegative off-diagonal elements at any $\xi \in \mathbb{K}$.

A particular case is the one in which $J_f(\xi)$ is also an irreducible matrix; i.e. $J_f(\xi)$ cannot be put in the following form

$$\left(\begin{array}{c|c} A & B \\ \hline 0 & C \end{array} \right)$$

by any reordering of the components of ξ. Somewhat inaccurately, $J_f(\xi)$ is irreducible for all ξ if system (4.1) cannot be decomposed into two subsystems, one of which does not depend on the other (think of linear systems with a coefficient matrix with the form shown above).

Condition (4.5) expresses in mathematical terms the concept of cooperativity of the system (or positive feedback).

A graphical representation of a cooperative system of two interacting species (groups, populations,...) is given in Fig. 4.1 where we have ignored the internal behavior of each species.

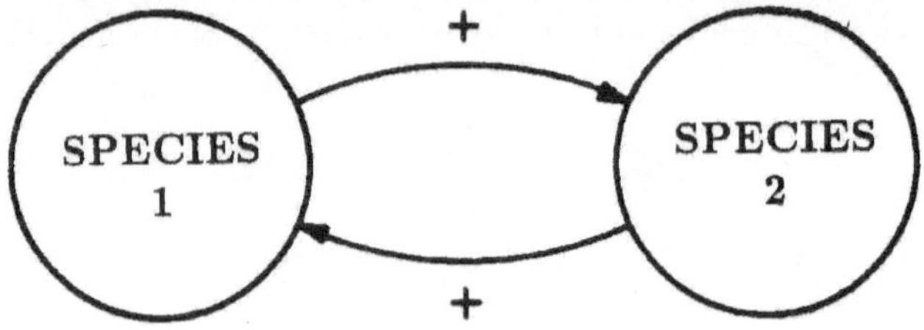

Fig. 4.1.

For a more extensive treatment of these concepts from a mathematical point of view we may refer to [203] and its references. From the biological point of view see [77].

We shall list now a series of epidemic models which fall in the class of quasimonotone systems, as defined in this section.

4.3. Epidemic models with a positive feedback

We shall list here classical epidemic models which show a quasimonotone structure. Typically these are two population systems which interact as depicted in Fig. 4.1. The systems have been illustrated here in their very basic formulation to show the positive feedback kind of interaction. In the subsequent mathematical analysis we shall refer to the simplest of these models, still to eliminate technical complications, and shall grow out of that to include more realistic features of the models, such as seasonal variation of the parameters, space structure, etc.

4.3.1. Gonorrhea [71, 73]

This model has already been presented in 2.2.1. We shall consider here its reduced form (2.17) that we rewrite here as

$$(4.6)\qquad \begin{cases} \dfrac{dI_1}{dt} = \alpha_{12} I_2 (1 - I_1) - \mu_1 I_1 \\ \dfrac{dI_2}{dt} = \alpha_{21} I_1 (1 - I_2) - \mu_2 I_2 \end{cases}$$

It is clear that system (4.6) has the quasimonotone structure (4.4). The threshold parameter is

$$(4.7)\qquad \rho = \frac{\alpha_{12}\alpha_{21}}{\mu_1 \mu_2}$$

such that if $\rho > 1$ a nontrivial endemic state appears, with components

$$I_1^* = \frac{\alpha_{12}\alpha_{21} - \mu_1\mu_2}{\mu_1\alpha_{21} + \alpha_{12}\alpha_{21}} \quad ; \quad I_2^* = \frac{\alpha_{12}\alpha_{21} - \mu_1\mu_2}{\mu_2\alpha_{12} + \alpha_{12}\alpha_{21}}$$

4.3.2. Schistosomiasis [126]

We shall refer to Macdonald 's model as reported in [159, 23] (see also [22, 89, 175]).

If we denote by X the current number of egg laying schistosomes in the human host population, and by Y the current number of infected snails in the environment, the simplest mathematical model is based on the following ODE system,

$$\begin{cases} \frac{dX}{dt} = \alpha\sigma Y - \gamma X \\ \frac{dY}{dt} = g(X)\left(1 - \frac{Y}{N}\right) - \delta Y \end{cases} \tag{4.8}$$

Here σ is the human population density per unit accessible water area; α is a multiplication rate due to the infected snail population; $g(X)$ is the force of infection acting on the uninfected snail population (N is the constant total population of snails), due to X; γ and δ are the intrinsic death rates of the two populations X and Y respectively.

The simplest choice for the force of infection, as reported in [126], is the linear one

$$g(X) = \beta X. \tag{4.9}$$

But, if we take into account the sexual behavior of schistosomes, $g(X)$ has to be chosen as a nonlinear function of X having the following properties [95].

$$\begin{aligned} 0 < g(X) < X, &\qquad X > 0 \\ g'(X) > 0, &\qquad X > 0 \\ g''(X) > 0, &\qquad X > 0 \\ g(X) \sim \frac{1}{2}X^2, &\qquad X \to 0 \\ g(X) \sim X - \sqrt{\frac{2}{\pi}}X, &\qquad X \to \infty \end{aligned} \tag{4.10}$$

In either cases we may see that the condition of quasimonotonicity (4.4) is satisfied.

System (4.8) clearly is a positive feedback system as illustrated in Fig. 4.1.

In fact it can be considered as an example of man-environment-man disease, the snail population acting as the environment intermediate host.

In the linear case (4.9) the epidemic system (4.8) admits in addition to the trivial equilibrium a nontrivial endemic equilibrium

$$X^* = N\left(\frac{\alpha\sigma}{\gamma} - \frac{\delta}{\beta}\right); \quad Y^* = N\left(1 - \frac{1}{R}\right) \tag{4.11}$$

whenever

$$R := \frac{\alpha\beta\sigma}{\gamma\delta} > 1 \quad ,$$

otherwise the only equilibrium is the trivial solution.

Thus the endemic situation can be eliminated by reducing either α or β or both. In fact sanitation is aimed at reducing β; safe water supplies at reducing α, though it may also have some benefits through β.

4.3.3. The Ross malaria model [14, 173]

The earliest attempt to provide a quantitative understanding of the dynamics of malaria transmission is the so called Ross-Macdonald model, which still is the basis for much malarial epidemiology [191].

This model, which captures the basic features of the interaction between the infected proportions of the human host and the mosquito vector population, is defined as follows :

$$\begin{cases} \dfrac{dx}{dt} = \alpha y(1-x) - rx \\ \dfrac{dy}{dt} = \beta x(1-y) - \mu y \end{cases} \tag{4.12}$$

where

x is the proportion of the human infected population;
y is the proportion of the female mosquito infected population;
α is the rate of infection of humans by mosquitoes;
β is the rate of infection of mosquitoes by humans;

$\frac{1}{r}$ is the average duration of infection in the human host;

$\dfrac{1}{\mu}$ is the average lifetime of a mosquito.

In this simple model, the total population of both humans and mosquitoes is assumed to be constant.

Further discussion can be found in [14].

It is easily seen that model (4.12) has exactly the same mathematical structure of the basic model for schistosomiasis (4.8) when the linear choice (4.9) is made for the force of infection.

Also it has the same structure of the gonorrhea model as reduced in (2.17).

It is clear then that the quasimonotonicity condition (4.4) is satisfied.

Usually the parameter α is explicited as follows

$$\alpha = \frac{a\beta M}{N} \tag{4.13}$$

where β is defined as above; a is the proportion of infected bites on man that produce an infection; N is the size of the human population; and M is the size of the female mosquito population.

The threshold parameter usually referred to is the basic reproductive rate [9] (better called basic reproduction ratio according to Diekmann et al. [79])

$$R = \frac{M}{N}\frac{\beta^2 \alpha}{\mu r} \; . \tag{4.14}$$

An interesting discussion about the meaning of the reproduction number in deterministic models of infectious diseases can be found in [124].

When $R > 1$ a unique nontrivial endemic state appears with components

$$x^* = \frac{R-1}{R+\dfrac{\beta}{\mu}}; \qquad y^* = \left(\frac{R-1}{R}\right)\left(\frac{\dfrac{\beta}{\mu}}{1+\dfrac{\beta}{\mu}}\right) . \tag{4.15}$$

4.3.4. A man-environment-man epidemic system

A mathematical model was proposed in [60] to describe the evolution of fecal-oral transmitted diseases in the Mediterranean regions (typhoid fever, cholera, infectious hepatitis).

This model includes as a basic feature, the positive feedback interaction between the infective human population and the concentration of bacteria (or viruses) in the environment, according to the scheme in Fig. 4.1.

The human population, once infected, acts as a multiplicator of the infectious agent which is then sent to the environment via fecal excretion; on the other hand the infectious agent is transmitted to the human population via contaminated food consumption.

If we assume that the total human population is very large with respect to its infected fraction, the basic mathematical model [60] can be written as follows:

$$\begin{cases} \dfrac{dz_1}{dt} = -a_{11}z_1 + a_{12}z_2 \\ \dfrac{dz_2}{dt} = -a_{22}z_2 + g(z_1) \end{cases} \tag{4.16}$$

where

- z_1 denotes the concentration of the infectious agent in the environment;
- z_2 denotes the human infective population;
- a_{ii} the intrinsic decay rates of the i-th population ($i = 1, 2$).
- a_{12} the multiplication rate of the infectious agent due to the human infected population;
- $g(z_1)$ the force of infection of the human population due to the concentration z_1 of infectious agent.

Differential systems of this form have already been used as mathematical models of different biological systems [199, 100, 21].

For the qualitative behavior of the system it is very relevant the shape of g as a function of z_1.

If we assume that g is an increasing function of z_1, then clearly system (4.16) is a quasimonotone system.

In fact, due to the very simplified mathematical structure of (4.16) we shall refer to it as a typical example of quasimonotone systems (see Section 4.4 and the following ones).

4.4. Qualitative analysis of the space-homogeneous autonomous case

As we have seen in Section 4.3 , for positive feedback epidemic systems, typically the internal behavior of each compartment is such that the population tends to extinction, so that only the positive feedback from other compartments is responsible for sustaining the possible nontrivial endemic state.

In order to clarify the analysis we shall limit ourselves to consider "man-environment-man" epidemics as a case study, the other models being an easy extension of that case (multigroup models will be analyzed in Sect. 4.6).

In the ODE case then we shall consider two dimensional systems in which z_1 represents the concentration of the "infectious agent" which pollutes the environment; z_2 represents the "human" population which is infected by z_1 and as a consequence acts as a multiplier of the infectious agent.

We shall assume as announced before that the scheme of interaction of the two "species" is the following

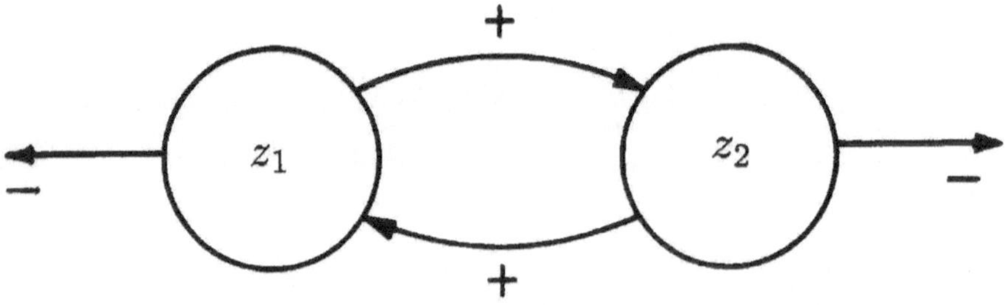

Fig. 4.2.

In mathematical terms we shall consider ODE systems of the form:

$$(4.17) \qquad \begin{cases} \dfrac{dz_1}{dt} = -a_{11}z_1 + g_1(z_2) \\ \dfrac{dz_2}{dt} = -a_{22}z_2 + g_2(z_1) \end{cases}$$

To reduce further technical complications which may only obscure the methods of analysis we shall assume that g_1 is linear

$$g_1(z_2) = a_{12}z_2$$

so that we may reduce ourselves to analyze

$$
\begin{cases} \dfrac{dz_1}{dt} = -a_{11}z_1 + a_{12}z_2 \\ \dfrac{dz_2}{dt} = -a_{22}z_2 + g(z_1) \end{cases} \tag{4.18}
$$

We shall see that by choosing g in different suitable ways we will have different dynamical behaviors of the system.

The function $g(z_1)$ plays the same role of the "force of infection" in the Kermack-McKendrick like models.

In the context of ecological modelling $g(z_1)$ is usually called the "functional response" of species (2) due to species (1) [120, 90].

We shall limit ourselves to two specific cases illustrated in Fig. 4.3; the case with constant concavity (a) and the case of a sigma-type functional response.

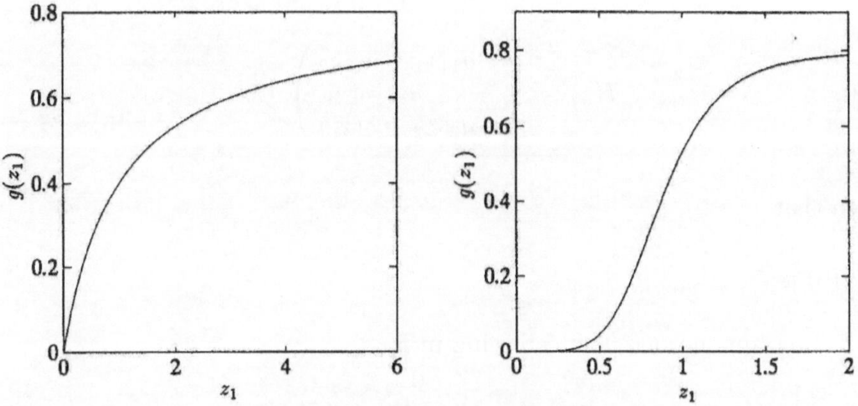

Fig. 4.3.

These two cases, it is the opinion of the author, contain most of the features of forces of infection in real epidemics (see also the discussion in Sect. 3).

Case (a) includes, for small values of z_1 the linear force of infection of the Kermack-McKendrick like models (law of mass action); while case (b) includes the possibility of a reaction of the human body to small amounts of acquired pathogenic material.

Both functions, for large values of z_1, tend to saturation.

This is not crucial in the analysis, but has a reasonable biological support based on the fact that actually $g(z_1)$ is related to a probability of infection that cannot grow indefinitely even if z_1 grows [60] (see also [111]).

If we refer to the models presented in Section 4.3, the case in Fig. 4.3 includes the model for schistosomiasis [89, 175, 4].

We shall make in the sequel the following assumptions on the function g.

(i) $g : \mathbb{R}_+ \to \mathbb{R}_+$ is a twice continuously differentiable function, such that $0 \leq g'(0+) < \infty$;

(ii) if $0 < z' < z''$, then $0 < g(z') < g(z'')$;

(iii) $g(0) = 0$;

(iv) $\limsup_{z \to +\infty} \dfrac{g(z)}{z} < \dfrac{a_{11}a_{22}}{a_{12}}$.

As a consequence of conditions (i)-(iv) the vector field

$$f(z) := \begin{pmatrix} -a_{11}z_1 + a_{12}z_2 \\ \\ -a_{22}z_2 + g(z_1) \end{pmatrix} \tag{4.19}$$

is such that

(F1) $f(0) = 0$

(F2) f is quasimonotone nondecreasing in $\mathbb{K}$

(F3) for any $z \in \mathbb{K}$, a $z_o \in \overset{\circ}{\mathbb{K}}$ exists, s.t. $z \leq z_o$ and $f(z_o) \ll 0$

Here $\mathbb{K} := \mathbb{R}_+ \times \mathbb{R}_+$.

Thanks to (F3) a bounded invariant rectangle can be defined for any solution of system (4.18) subject to an initial condition in $\mathbb{K}$, and this implies that such a solution can be extended to the whole $\mathbb{R}_+$.

Given the assumptions (i)-(iv) we shall distinguish two main cases:

(v) $0 < g'(0+) < +\infty$; $g''(z) < 0$ for any $z > 0$ (Fig. 4.3(a))

and

(v') a $\tilde{z} > 0$ exists s.t. $\begin{array}{ll} g''(z) > 0 & \text{for } 0 < z < \tilde{z} \\ g''(z) < 0 & \text{for } z > \tilde{z} \end{array}$ (Fig. 4.3(b))

Let us analyze case (v) first. In this case the system of the two isoclines

$$(4.19)' \quad \begin{cases} f_1(z_1,z_2) := -a_{11}z_1 + a_{12}z_2 = 0 \\ f_2(z_1,z_2) := -a_{22}z_2 + g(z_1) = 0 \end{cases}$$

have one trivial and at most one nontrivial solution depending on the value of

$$(4.20) \quad \theta := \frac{a_{12}\, g'(0)}{a_{11}\, a_{22}}$$

It is in fact easy to show the following

Proposition 4.1. *Under conditions (i)-(v),*

a) *If* $0 < \theta \leq 1$ *then system (4.18) admits only the trivial solution* $0 = (0,0)'$ *as an equilibrium solution.*
b) *If* $\theta > 1$ *then system (4.18) admits two equilibrium solutions;* 0 *and another nontrivial solution* $z^* \in \overset{\circ}{\mathbb{K}}$.

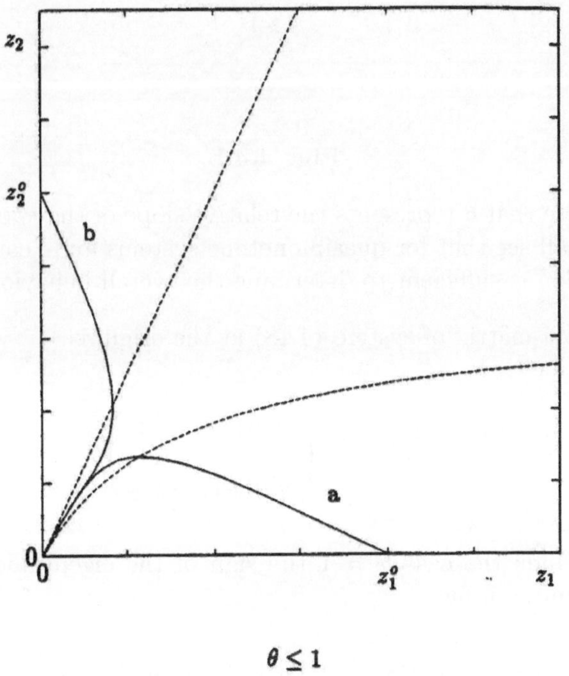

Fig. 4.4.a.

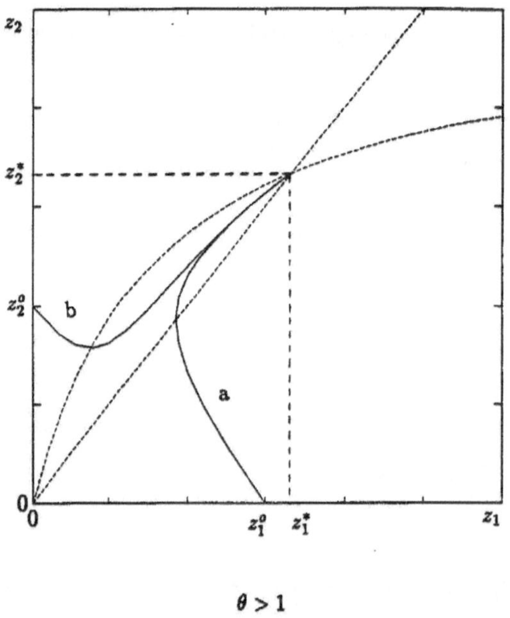

Fig. 4.4.b.

Remark. Note that θ represents the relative slope of the two isoclines at the origin. We shall see that for quasimonotone systems and "concave" isoclines, this information is sufficient to determine the overall behavior of the system.

The Jacobi matrix of system (4.18) at the origin is

$$J_f(0+) = \begin{pmatrix} -a_{11} & a_{12} \\ g'(0+) & -a_{22} \end{pmatrix}$$

If we exclude the case $\theta = 1$ the sign of the eigenvalues λ_1 and λ_2 of $J_f(0+)$ is definite; in fact

$$\begin{aligned} \theta < 1 &\implies \lambda_1, \lambda_2 < 0 \qquad \text{(stable node)} \\ \theta > 1 &\implies \lambda_1 > 0, \lambda_2 < 0 \qquad \text{(unstable saddle)} \end{aligned}$$

while

$$\theta = 1 \quad \Longrightarrow \quad \lambda_1 < 0, \lambda_2 = 0$$

Thus the linear analysis in the case $\theta = 1$ does not provide sufficient information.

It is clear on the other hand that due to the above information, if $\theta < 1$ then the trivial solution is LAS while for $\theta > 1$ it is unstable.

Global methods based on the technique of nested contracting rectangles (see Appendix A, Section A.4.2.1) lead to the following

Theorem 4.2. *Suppose conditions (i)-(v) are satisfied.*

a) If $0 < \theta \leq 1$ *then 0 is GAS for system (4.18) in* $\mathbb{K} := \mathbb{R}_+ \times \mathbb{R}_+$.
b) If $\theta > 1$ *then 0 is unstable and* z^* *is GAS for system (4.18) in* $\dot{\mathbb{K}} = \mathbb{K} - \{0\}$.

Proof. As announced, the proof is based on the nested contracting rectangles technique.

Let us describe this technique to show the GAS of z^* in $\dot{\mathbb{K}}$.

Translate first the coordinate system (z_1, z_2) into a new coordinate system (U_1, U_2) such that z^* is translated into the origin $\bar{\mathrm{O}}$ of this new system. Denote by $F(U)$ the translated vector field

$$F(U) := \begin{pmatrix} -a_{11}U_1 + a_{12}U_2 \\ \\ G(U_1) - a_{22}U_2 \end{pmatrix} \tag{4.21}$$

where

$$G(U_1) = g(U_1 + z_1) - g(z_1^*)$$

Let us denote by (see Fig. 4.5)

$$\mathcal{U} := \left\{ (U_1, U_2) \in \mathbb{R}_- \times \mathbb{R}_- - \{0\} \;\middle|\; \frac{a_{11}}{a_{12}} U_1 < U_2 < \frac{1}{a_{22}} G(U_1) \right\} \tag{4.22}$$

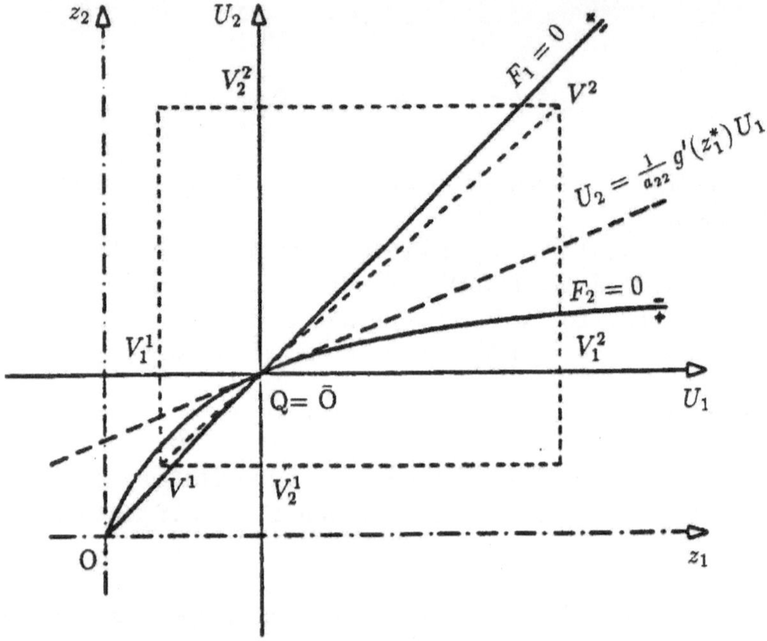

Fig. 4.5.

It can be easily seen that if $V^1 := \left(V_1^1, V_2^1\right)^T \in \mathcal{U}$ and $V^2 := \left(V_1^2, V_2^2\right)^T$ is an arbitrary point on the straight line passing through V^1 and $\bar{O}$, with $V_1^2, V_2^2 > 0$, the rectangle

$$\mathcal{R} := \left[V_1^1, V_1^2\right] \times \left[V_2^1, V_2^2\right]$$

is contracting for $F(U)$. Furthermore, for any $\tau \in (0,1]$ the rectangle $\tau\mathcal{R}$ is contracting for $F(U)$ [56].

By applying Theorem A.39 in Appendix A, Section A.4.2.1, the theorem follows.

Consider now the case (v') : the following theorem holds [58].

Theorem 4.3.

a) If $\theta < 1$ *, and for any* $z_1 > 0$,

(4.23)
$$\frac{g(z_1)}{z_1} < \frac{a_{11}a_{22}}{a_{12}}$$

then the trivial solution is GAS in the whole $\mathbb{K}$ *for system (4.23).*

b) *If* $\theta > 1$ *then system (4.18) admits, in addition to the origin O, only one nontrivial solution* $z^* \in \overset{\circ}{\mathbb{K}}$. *In this case O is unstable while* z^* *is GAS in* $\dot{\mathbb{K}}$.

c) *If* $\theta < 1$ *and for some* $z_1 > 0 : \dfrac{g(z_1)}{z_1} > \dfrac{a_{11}a_{22}}{a_{12}}$ *then system (4.18) admits three equilibrium points in the positive quadrant* $\mathbb{K}$*; the origin O and two other nontrivial equilibria* z^* *and* z^{**}. *They are such that* $0 \ll z^* \ll z^{**}$. *In this case* z^* *is a saddle point; if we denote by* M_+ *the stable manifold of* z^*, *it divides the cone* $\mathbb{K}$ *in two regions; the domain of attraction of O, say dom(O), and the domain of attraction of* z^{**}, *say dom*(z^{**}) *:*

$$\mathbb{K} = M_+ \cup dom(O) \cup dom(z^{**})$$

the three sets being disjoint subsets of $\mathbb{K}$ *(note that* M_+ *is the domain of attraction of* z^**). O and* z^{**} *are GAS in their own domain of attraction.*

Proof.

a) Let B_1 denote the matrix

(4.24)
$$B_1 := \begin{pmatrix} -a_{11} & a_{12} \\ \bar{a}_{21} & -a_{22} \end{pmatrix}$$

where $\bar{a}_{21} := \sup_{z \in \mathbb{R}_+} \dfrac{g(z_1)}{z_1}$. Then

(4.25)
$$f(z) \leq B_1 z \qquad \text{for any} \qquad z \in \mathbb{K}.$$

Due to condition (4.23) B_1 has both negative eigenvalues. Case a) of the theorem follows by the comparison theorem (Theorem A.32 in Appendix A, Section A.4.2).

b) Clearly, under condition $\theta > 1$, system (4.18) admits two equilibrium points; O and $z^* \in \overset{\circ}{\mathbb{K}}$. It is easy to build a family of nested contracting rectangles around z^*, thus implying the GAS of z^* in $\dot{\mathbb{K}}$ (see Fig. 4.6).

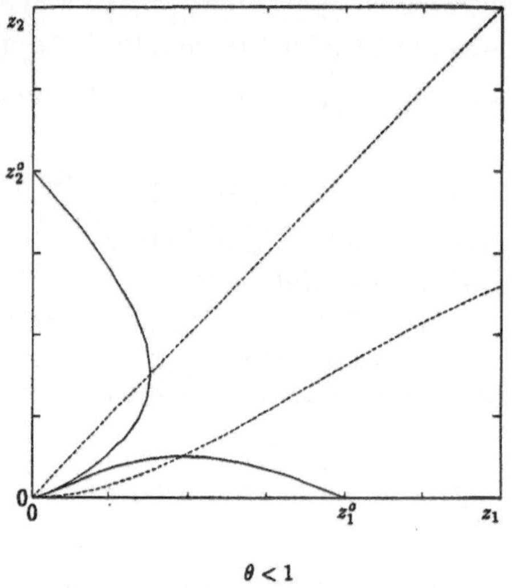

Fig. 4.6.a. Phase plane portrait of the solutions. Dotted lines (- - -) show the isoclines.

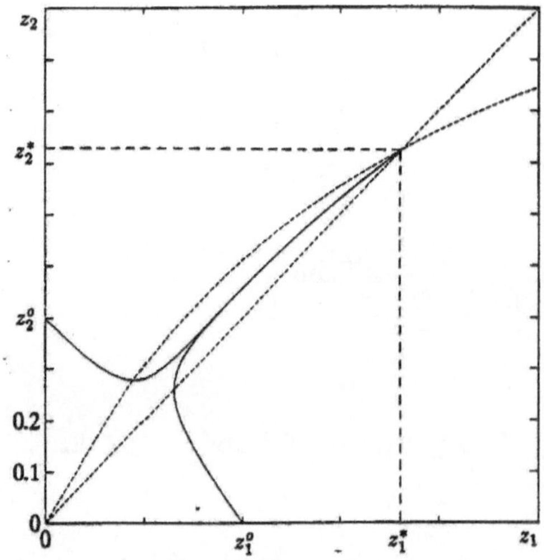

Fig. 4.6.b. Phase plane portrait of the solutions. Dotted lines (- - -) show the isoclines.

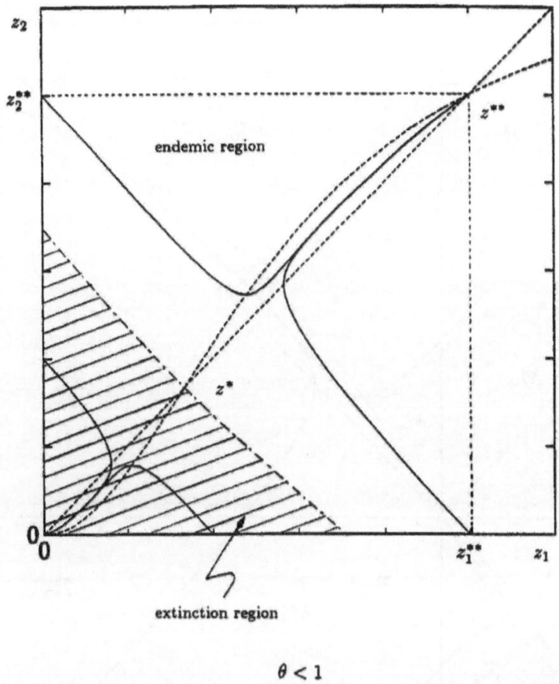

Fig. 4.6.c. Phase plane portrait of the solutions. Dotted lines (- - -) show the isoclines.

c) The Jacobi matrix of f at $z \in \mathbb{K}$ is given by

$$J_f(z) = \begin{pmatrix} -a_{11} & a_{12} \\ g'(z_1) & -a_{22} \end{pmatrix}$$

Under the assumptions of case c) $J_f(0+)$ and $J_f(z^{**})$ have both negative eigenvalues; hence O and z^{**} are LAS for system (4.18). On the other hand $J_f(z^*)$ has one positive and one negative eigenvalues; it is then a saddle point. Due to well known results [104] it can be shown that in $\mathbb{K}$ there exist two curves M_+ and M_- respectively known as the stable manifold and unstable manifold of z^*. M_+ is positively invariant and for any $z \in M_+ : \lim_{z \to +\infty} V(t)z = z^*$. M_- is negatively invariant and for any $z \in M_- : \lim_{t \to -\infty} V(t)z = z^*$.

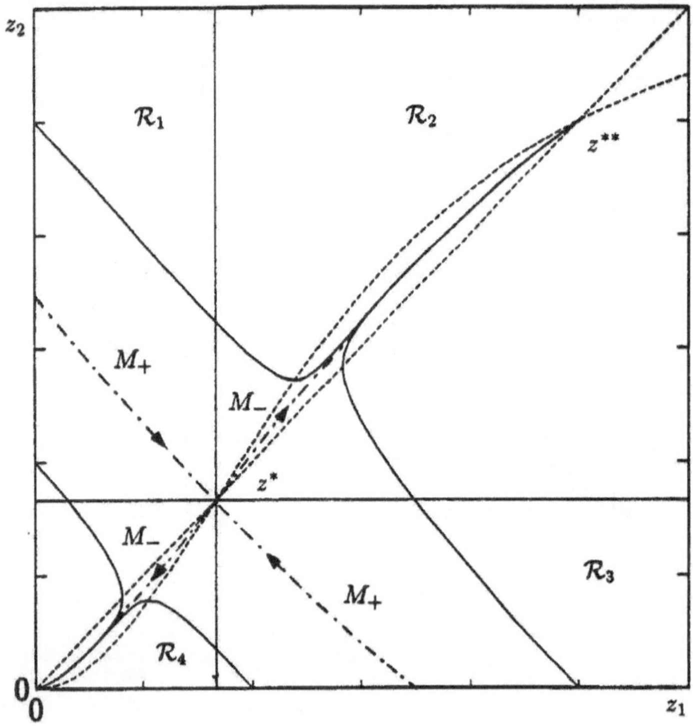

Fig. 4.7.

Arguments based on nested contracting rectangles show that O is GAS in $\mathcal{R}_4 := \{z \in \mathbb{K} \mid 0 \leq z \ll z^*\}$ and z^{**} is GAS in $\mathcal{R}_2 := \{z \in \mathbb{K} \mid z^* \ll z\}$ (see Fig.4.7).

Monotone techniques based on the fact that f is quasimonotone imply that O is GAS in dom(O) while z^{**} is GAS in dom(z^{**}) (see [199]).

We wish to remark here that the parameter θ as defined in (4.20) does not play any more by itself the role of a "threshold parameter".

In fact while it is still true that if $\theta > 1$ a nontrivial endemic level exists which is GAS in the feasible cone $\mathbb{K}$, when $\theta < 1$ we need to distinguish two subcases; one in which the epidemic always tends to extinction, and another one in which the epidemic tends to extinction only if the initial condition is below a separatrix M_+ (not completely specified analytically); otherwise it tends to a nontrivial endemic state.

This case is of particular epidemiological interest in that it may explain why only major perturbations with respect to the trivial steady state may give rise to endemic situations; in this case in fact this "threshold effect" is induced by a functional response $g(z_1)$ of the force of infection for which $g'(0) = 0$ because of a possible reaction of the host antibodies to the infectious agent.

An extension to the space heterogeneous case can be found in Sect. 5.4.

4.5. The periodic case

It may happen that the parameters of system (4.18) depend on time in a periodic way because of seasonal variations, having the same period $\omega \geq 0$.

We shall show in this section that whenever this happens, nontrivial periodic solutions may appear with the same period ω of the parameters.

Again this strongly depends upon the quasimonotone structure of the system. For simplicity we shall assume that only the nonlinear force of infection depends on time in a periodic way, with period $\omega \geq 0$:

$$g(t; z_1) = p(t)h(z_1) \tag{4.26}$$

with

$$p(t+\omega) = p(t) \ , \qquad t \in \mathbb{R} \tag{4.27}$$

So that system (4.18) will be rewritten as

$$\frac{dz}{dt} = f(t; z) \ , \qquad t > 0 \tag{4.28}$$

with

$$f(t; z) = \begin{pmatrix} -a_{11}z_1 + a_{12}z_2 \\ \\ -a_{22}z_2 + p(t)h(z_1) \end{pmatrix} , \qquad z \in \mathbb{K} \tag{4.29}$$

We define

$$p_{\min} := \min_{t \in [0,\omega]} p(t); \qquad p_{\max} := \max_{t \in [0,\omega]} p(t) \tag{4.30}$$

and assume that h satisfies the following properties

(i) $h : \mathbb{R}_+ \to \mathbb{R}_+$ is a twice continuously differentiable function;

(ii) if $0 < z' < z''$ then $0 < h(z') < h(z'')$

(iii) $h(0) = 0$

(iv) $\limsup_{z \to +\infty} \frac{h(z)}{z} < \frac{a_{11}a_{22}}{a_{12}p_{\max}}$

(v) $0 < h'(0+) < +\infty;\quad h''(z) < 0$ for any $z > 0$

Due to the assumptions on h we can state the following properties for f:

(F1) for any $t \in \mathbb{R}: \quad f(t;0) = 0$
(F2) for any $t \in \mathbb{R}: \quad f(t;\cdot)$ is a quasimonotone nondecreasing function in $\mathbb{K}$;
(F3) for any $z \in \mathbb{K}$, there exists a $z_o \in \mathring{\mathbb{K}}$ s.t. $z \le z_o$ and $f(t;z_o) \ll 0$.
(F4) for any $t \in \mathbb{R}: \quad f(t;\cdot) = f(t+\omega;\cdot)$

Under these properties it can be shown that a unique solution exists for problem (4.28) subject to any initial condition

$$z(s) = z^o \in \mathbb{K}, \quad s \in \mathbb{R} \tag{4.28s}$$

Due to (F3) it can be extended to the whole $[s, +\infty)$ [139].

For any $(s,t) \in \Sigma := \{(s,t) \in \mathbb{R}^2 \quad | \quad s \le t\}$, we shall denote by $V(t,s)z(s)$ the unique solution of problem (4.28), (4.28s).

We shall define monodromy operator or Poincaré map of system (4.28) the following

$$V = V(\omega;0) \ . \tag{4.31}$$

In this case the Jacobi matrix of f at any $z \in \mathbb{K}$ is given by:

$$J_f(t;z) = \begin{pmatrix} -a_{11} & a_{12} \\ p(t)h'(z) & -a_{22} \end{pmatrix}, \quad t \in \mathbb{R}, \ z \in \mathbb{K} \tag{4.32}$$

By assuming that g satisfies assumptions (i)-(v), it is easily seen that the following lemma holds.

Lemma 4.4. *For any* $t \in \mathbb{R}$ *, and any* $z \in \mathring{\mathbb{K}}$, $J_f(t;z)$ *is a quasi-monotone increasing matrix;* $zJ_f(t;z) - f(t;z)$ *has nonpositive components and one of them is negative, for any* $t \in \mathbb{R}$, $z \in \mathring{\mathbb{K}}$; *and finally f satisfies properties (a) and (b) in [139; Lemmas 10.1, 10.4].*

Thanks to Lemma 4.4, the following theorem holds [59].

Theorem 4.5. *If*

$$\theta^{\max} := \frac{a_{12} p_{\max} h'(0+)}{a_{11} a_{22}} < 1 \tag{4.33}$$

then the trivial solution is GAS in $\mathbb{K}$ for system (4.28). If

$$\theta^{\min} := \frac{a_{12} p_{\min} h'(0+)}{a_{11} a_{22}} > 1 \tag{4.34}$$

then a unique nontrivial periodic solution (with positive components) exists, which is GAS in $\dot{\mathbb{K}} := \mathbb{K} - \{0\}$ *for system (4.28).*

Proof. Let

$$B_{\max} := \begin{pmatrix} -a_{11} & a_{12} \\ p_{\max} h'(0+) & -a_{22} \end{pmatrix} \tag{4.35}$$

Thanks to the basic assumptions on h,

$$\text{for any} \quad t \in \mathbb{R}, \ \text{for any} \quad z \in \mathbb{K}: \quad f(t;z) \le B_{\max} z \tag{4.36}$$

It can be easily seen that condition (4.33) implies that the eigenvalues of matrix $B_{\max}$ are both negative, from which the first part of the theorem follows.

As far as the second part is concerned, according to [135] since $f(t;z)$ satisfies Lemma 4.4, we only need to show that conditions for the increase of small solutions and for the decrease of large solutions are also fulfilled here.

We start by observing that if we define

$$B_{\min} := \begin{pmatrix} -a_{11} & a_{12} \\ p_{\min} h'(0+) & -a_{22} \end{pmatrix} \tag{4.37}$$

then, thanks to condition (4.34), $B_{\min}$ has a positive eigenvalue. The same holds for the matrix

$$B^{\epsilon}_{\min} := \begin{pmatrix} -a_{11} & a_{12} \\ p_{\min} h'(0+) - \epsilon & -a_{22} \end{pmatrix} \tag{4.38}$$

for a suitable choice of $\epsilon > 0$. It is clear that a $\delta > 0$ exists such that

$$\text{for any} \quad t \in \mathbb{R} \quad \text{and} \quad z \in \mathbb{K}, \quad |z| \le \delta : f(t; z) \ge B^{\epsilon}_{\min} z. \tag{4.39}$$

This implies the required increase of "small" solutions.
On the other hand, if we choose $a_{21} \in \mathbb{R}$ such that

$$\limsup_{z \to +\infty} \frac{h(z)}{z} < a_{21} \frac{a_{11}\, a_{22}}{a_{12}\, p_{\max}} \tag{4.40}$$

(see assumption (iv) on h), and consider the matrix

$$C := \begin{pmatrix} -a_{11} & a_{12} \\ a_{21} & -a_{22} \end{pmatrix} \tag{4.41}$$

then both the eigenvalues of C are negative and a $\delta > 0$ exists such that

$$\text{for any} \quad t \in \mathbb{R} \quad \text{and} \quad z \in \mathbb{K}, \quad |z| \ge \delta : f(t; z) \le C z\ . \tag{4.42}$$

this implies the decrease of "large" solutions, and completes the proof.

We wish to remark that Theorem 4.5 only gives sufficient conditions for the existence and the GAS of the nontrivial periodic endemic state. Similarly condition (4.33) is only a sufficient "threshold" condition for the GAS of the trivial solution.

The results contained in Theorem 4.5 can be obtained in different ways (see [18, 202]) but it has been our explicit intention to refer to the fundamental work of Krasnoselskii and his School [138, 139] which contains most of the key ideas of subsequent work. We wish also to quote the important work done more recently by M. Hirsch who gives a systematic account of the theory of monotone flows in topological spaces [119].

The work by R. Martin [161] has also to be quoted, but to it we shall refer better in the context of epidemic systems with spatial diffusion.

4.6. Multigroup models

In this section we extend our methods to the analysis of multigroup quasi-monotone system, as a first (discrete) case of nonhomogeneous populations.

4.6.1. A model for gonorrhea in a nonhomogeneous population [142]

In the model presented in [142] an SIS model is confirmed by Lajmanovich and Yorke for gonorrhea transmission; the sexually active population is divided into two main classes, the susceptible class S and the infective class I such that it is assumed that the total active population $N(t) = S(t) + I(t)$ is a constant at any $t \geq 0$.

For a nonhomogeneous population Lajmanovich and Yorke propose the model already presented in Sect. 2.3.4.1, system (2.92), where we announced that monotone techniques would have been more suitable to analyze the asymptotic properties of the system.

In this presentation we shall follow our own framework, based on the results of Appendix A, Section A.4 .

Let us rewrite system (2.92) for our convenience

$$\text{(4.43)} \qquad \begin{cases} \dfrac{d}{dt} I_i = \left(\displaystyle\sum_{j=1}^{n} \lambda_{ij} I_j \right) (1 - I_i) - \gamma_i I_i - \mu_i I_i \\ I_i + S_i = 1 \end{cases}$$

for $i = 1, \ldots, n$.

Let $z := (I_1, \ldots, I_n)^T$ and let

(4.44) $$f : z = (I_1, \ldots, I_n)^T \in \mathbb{R}^n \to f(z) = (f_i(z))_{1\le i\le n} \in \mathbb{R}^n$$

such that

(4.45) $$f_i(z) := \left(\sum_{j=1}^{n} \lambda_{ij} I_j \right) (1 - I_i) - (\gamma_i + \mu_i)\, I_i \ .$$

System (4.43) can be rewritten as

(4.46) $$\frac{dz}{dt} = f(z) \ , \qquad t > 0 \ .$$

It is easily seen that the initial value problem associated with (4.43) is well posed for any choice of the initial condition

(4.47) $$z^o \in \mathcal{R} := \ [0, 1]^n \subset \mathbb{R}^n \ ,$$

i.e. a unique solution $z(t)$ exists for any time $t \geq 0$. The existence in the large is a consequence of Lemma 4.6 .

Clearly f is quasimonotone increasing with respect to the cone $\mathbb{K} := \mathbb{R}^n_+$ (see Sect. 4.2 for the definitions); in fact the off-diagonal terms of the Jacobi matrix , $J_f(z)$, of f at any $z \in \mathcal{R}$ are given by

(4.48) $$(J_f(z))_{ij} = \frac{\partial f_i}{\partial I_j} = (1 - I_i)\, \lambda_{ij} \geq 0 \ , \qquad i \neq j \ , \quad i, j = 1, \ldots, n \ .$$

It is assumed that $\gamma_i + \mu_i > 0 \ , \ i = 1, \ldots, n$.

As a consequence of the above we can state that

(F1) $f(0) = 0$

(F2) f is quasimonotone nondecreasing in $\mathcal{R}$.

(F3) If we let $\mathbf{1} := (1, \ldots, 1)^T \in \mathcal{R}$ then $f(\mathbf{1}) \ll 0$.

Conditions (F1)-(F3) imply the corresponding conditions in Sect. 4.4 ; as a consequence the following lemma holds.

Lemma 4.6. *The rectangle $\mathcal{R}$ is a bounded invariant rectangle for any solution of system (4.43) subject to an initial condition in $\mathcal{R}$, and this implies that such a solution can be extended to the whole $\mathbb{R}_+$.*

Although λ_{ij} is not necessarily equal to λ_{ji} , it is assumed that

$$\lambda_{ij} \neq 0 \iff \lambda_{ji} \neq 0 \, . \tag{4.49}$$

It is assumed further that the matrix $\Lambda := (\lambda_{ij})_{i,j=1,\dots,n}$ is irreducible, so that the Jacobi matrix $J_f(z)$ will be so.

Further let us denote by

$$\partial\mathcal{R}_1 := \{z \in \mathcal{R} \mid z_i = 0 \quad \text{for some } i\}$$
$$\partial\mathcal{R}_2 := \{z \in \mathcal{R} \mid z_i = 1 \quad \text{for some } i\};$$

the boundary of $\mathcal{R}$ is given by $\partial\mathcal{R} = \partial\mathcal{R}_1 \cup \partial\mathcal{R}_2$. Consider now the outer normal η at a point $z \in \partial\mathcal{R}$.

On the face $z_i = 0$,

$$\eta = \eta_i^1 := (-1\,\delta_{ij})_{j=1,\dots,n} \qquad (i = 1, \dots, n) \, ;$$

on the face $z_i = 1$,

$$\eta = \eta_i^2 := (\delta_{ij})_{j=1,\dots,n} \qquad (i = 1, \dots, n) \, .$$

Now since

$$\left.\frac{dz}{dt}\right|_{z_i=0} \cdot \eta_i^1 = -\sum_{j=1}^{n} \lambda_{ij} z_j < 0$$

$$\left.\frac{dz}{dt}\right|_{z_i=1} \cdot \eta_i^2 = -(\gamma_i + \mu_i) < 0$$

the following lemma holds.

Let us denote, as in Sect. 4.2, the evolution operator of system (4.43) by $\{V(t), \ t \geq 0\}$ so that the solution of the system subject to the initial condition $z^o \in \mathcal{R}$ is given by $z(t) = V(t)\, z^o, \ t \geq 0$.

Lemma 4.7. *If Λ is irreducible, then*

$$z^o \in \dot{\mathcal{R}} \Longrightarrow V(t)\, z^o \in \mathring{\mathcal{R}}, \qquad \textit{for any} \quad t > 0 \tag{4.50}$$

($\dot{\mathcal{R}} := \mathcal{R} - \{0\}$ while $\mathring{\mathcal{R}}$ denotes, as usual, the interior of $\mathcal{R}$).

From Lemma 4.7, one can conclude that no group can have all the individuals infected for a positive time interval. Furthermore, no group can have all the individuals free of the disease for a positive time interval, unless everyone in all groups is uninfected. We wish to remark that this is a strict consequence of the irreducibility of the matrix of interactions Λ.

In particular we have shown that $V(t)$ is strictly positive.

Due to the quasimonotonicity of f in $\mathcal{R}$ (condition (F2)) we also have (Appendix A, Section A.4.2) that $\{V(t),\ t \in \mathbb{R}_+\}$ is a family of monotone nondecreasing operators in the (partially) ordered space $\mathbb{R}_+^n\ (\cap\mathcal{R})$:

$$\text{(4.51)} \quad \text{for any } z^1, z^2 \in \mathcal{R},\ z^1 \leq z^2 \Longrightarrow V(t)\,z^1 \leq V(t)\,z^2 \quad \text{for any } t \geq 0\,.$$

Now, the Jacobi matrix of system (4.43) at any $z \in \mathcal{R}$ is given by

$$\text{(4.52)} \qquad (J_f(z))_{ij} = \begin{cases} -(\gamma_i + \mu_i) + (1 - 2z_i)\,\lambda_{ii}\,, & j = i \\ (1 - z_i)\,\lambda_{ij}\,, & j \neq i \end{cases}$$

for $i, j = 1, \ldots, n$; so that for any $z \in \mathcal{R}$ we have that $J_f(z)$ is irreducible, and further it is monotone non increasing with respect to z in $\mathring{\mathcal{R}}$:

$$\text{(4.53)} \quad 0 < z' < z'' < \mathbf{1} \Longrightarrow (J_f(z'))_{ij} \geq (J_f(z''))_{ij}\,, \ \text{for } i, j = 1, \ldots, n\,.$$

All the above implies the following (see Appendix A, Section A.4.2)

(F4) for any $R > 0$ an irreducible quasimonotone increasing $n \times n$ matrix C_R exists such that

$$\text{for any } z', z'' \in \mathring{\mathcal{R}}, \quad 0 \leq z' \leq z'', \quad |z'|, |z''| \leq R :$$

$$f(z'') - f(z') \geq C_R\,(z'' - z')\ .$$

Lemma A.34 in Appendix A, Section A.4.2, implies the strict monotonicity of the evolution operator $\{V(t),\ t \in \mathbb{R}_+\}$:

$$\text{(4.54)} \quad \text{for any } z^1, z^2 \in \mathring{\mathcal{R}},\ z^1 \leq z^2,\ z^1 \neq z^2 :\ V(t)\,z^1 \ll V(t)\,z^2,\ t > 0\ .$$

From the definition (4.44), (4.45) the sublinearity of f follows

(F5) for any $z \in \overset{\circ}{\mathcal{R}}$, for any $\tau \in (0,1)$: $\quad f(\tau z) > \tau f(z)$.

As a consequence

$$\text{for any } z \in \overset{\circ}{\mathcal{R}}, \text{ for any } \tau \in (0,1): \quad f(z) < \frac{1}{\tau} f(\tau z)$$

and for $\tau \to 0+$:

$$\text{for any } \quad z \in \overset{\circ}{\mathcal{R}}: \quad f(z) \leq J_f(0)\, z \tag{4.55}$$

where

$$(J_f(0))_{ij} = \begin{cases} -(\gamma_i + \mu_i) + \lambda_{ii}\,, & i = j \\ \lambda_{ij}\,, & i \neq j \end{cases} \tag{4.56}$$

for $i, j = 1, \ldots, n$, is the Jacobi matrix of f at 0.

System (4.43) always admits the trivial equilibrium solution $\mathbf{0} := (0, \ldots, 0)^T$. Its stability properties are obtained as follows .

Consider first the linearized system associated to (4.43)

$$\frac{dv}{dt} = J_f(0)\, v\,, \qquad t > 0\;. \tag{4.57}$$

We shall denote by $\{V_L(t),\ t \geq 0\}$ the family of evolution operators of system (4.57).

As a consequence of $J_f(0)$ being a real quasimonotone nondecreasing irreducible matrix, we know (Perron-Frobenius Theorem; see Appendix A, Section A.4.1) that a dominant real eigenvalue μ exists equal to the spectral radius of the matrix $J_f(0)$.

The following theorem is a direct consequence of Theorem B.27 and Theorem B.28 in Appendix B.

Theorem 4.8. *Let $\mu \in \mathbb{R}$ be the dominant real eigenvalue of $J_f(0)$.*

a) If $\mu \leq 0$, then the trivial equilibrium is GAS in $\mathcal{R}$ for system (4.43).
b) If $\mu > 0$, then the trivial equilibrium is unstable for system (4.43). Further there exists a $k > 0$ such that for any $z^o \in \dot{\mathcal{R}}$:

$$\liminf_{t \to +\infty} |V(t)\, z^o| \geq k\;.$$

Under assumption b) of Theorem 4.8, due to the definition of f, an $\epsilon > 0$ can be chosen such that, $\epsilon < \min\{\lambda_{ij} \mid \lambda_{ij} > 0,\ i \neq j\}$, the matrix $J_f^\epsilon(0)$ defined as follows is still quasimonotone nondecreasing , irreducible and its leading eigenvalue $\mu^\epsilon \in \mathbb{R}$ is still $\mu^\epsilon > 0$:

$$\left(J_f^\epsilon(0)\right)_{ij} = \begin{cases} \left(J_f(0)\right)_{ii} & i = j \\ 0 & \text{if} \quad \lambda_{ij} = 0,\ i \neq j \\ \lambda_{ij} - \epsilon & \text{if} \quad \lambda_{ij} > 0,\ i \neq j \end{cases}$$

Correspondingly a $\delta_\epsilon > 0$ will exist such that

$$\text{(4.58)} \qquad \text{for any} \quad z \in \mathcal{R},\ |z| < \delta_\epsilon : \qquad f(z) \geq J_f^\epsilon(0)\, z \ .$$

Now, since $J_f^\epsilon(0)$ is a quasimonotone nondecreasing irreducible matrix, corresponding to its dominant real eigenvalue $\mu^\epsilon > 0$ an eigenvector $\eta \gg 0$ will exists (Perron-Frobenius Theorem; see Appendix A, Section A.4.1).

If $(\epsilon_k)_{k\in\mathbb{N}}$ is a decreasing infinitesimal sequence of positive real numbers $\epsilon_k \in (0,1)$ such that

$$\text{for any} \quad k \in \mathbb{N} : |\epsilon_k \eta| \leq \delta_\epsilon$$

(δ_ϵ is the one defined in (4.58)), then for any $k \in \mathbb{N}$, $\underline{z}^k := \epsilon_k \eta$ is a subsolution of system

$$\text{(4.59)} \qquad f(z) = 0$$

i.e. (see Appendix A, Section A.4.2.1)

$$\text{(4.60)} \qquad \text{for any} \quad k \in \mathbb{N} : \qquad f\left(\underline{z}^k\right) \geq 0$$

In fact, by (4.58), for any $k \in \mathbb{N}$,

$$f(\epsilon_k \eta) \geq J_f(0)\, \epsilon_k \eta = \epsilon_k \mu_\epsilon \eta \gg 0 \ .$$

Further, $\overline{z} := \mathbf{1}$ is a supersolution of system (4.59); in fact (Appendix A, Section A.4.2.1)

$$f(\overline{z}) \ll 0 \tag{4.61}$$

As a consequence (Lemma B.30 in Appendix B) $V(t)\,\overline{z}$ is monotone non-increasing in $t \geq 0$, and there exists a solution z_+^* of system (4.59) such that

$$\lim_{t \to +\infty} V(t)\,\overline{z} = z_+^*$$

Since, for any $z^o \in \overset{\circ}{\mathcal{R}} \, : \, z^o \leq \overline{z}$, by the comparison theorem (Appendix A, Section A.4.2) for any $t \geq 0 \, : \, 0 \leq V(t)\, z^o \leq V(t)\, \overline{z}$. In the limit

$$\lim_{t \to +\infty} d\left(V(t)\, z^o \, , \, [0, z_+^*]\right) = 0 \tag{4.62}$$

Since $J_f^\epsilon(0)$ is an irreducible quasimonotone increasing matrix, condition (4.58) together with Lemma 4.7 implies that provided z^* is a nontrivial equilibrium solution of system (4.43) it must be

$$z^* = V(t)\, z^* \gg 0 \, .$$

Now, since we are considering the case in which 0 is unstable, $z_+^* \neq 0$ so that it must be $z_+^* \gg 0$.

The existence of at least a nontrivial equilibrium solution let us now apply (Lemma B.29 Appendix B) to claim that for any $k \in \mathbb{N}$, $V(t)\, \underline{z}^k$ is monotone nondecreasing in $t \geq 0$, and further that exists an equilibrium solution $z_-^k \gg 0$ for system (4.43) i.e. a solution of (4.59), such that

$$\lim_{t \to +\infty} \left| V(t)\, \underline{z}^k - z_-^k \right| = 0 \, . \tag{4.63}$$

The sequence $(z_-^k)_{k \in \mathbb{N}}$ satisfies all the assumptions of Lemma B.31 in Appendix B so that an equilibrium solution $z_-^* \gg 0$ exists for system (4.43) such that

$$\lim_{k \to +\infty} z_-^k = z_-^* \;\; (= \inf_k \, z_-^k) \, .$$

Now, let $z^o \in \dot{\mathcal{R}}$; by Lemma 4.7 , if $\nu > 0$, we have $V(\nu)\, z^o \in \overset{\circ}{\mathcal{R}}$; so that for a sufficiently big $k \in \mathbb{N}$,

$$V(\nu)\, z^o \geq \epsilon_k \eta = \underline{z}^k \,.$$

As a consequence of the comparison theorem, for any $t > 0$ we have

$$V(t+\nu)\, z^o = V(t)\, V(\nu)\, z^o \geq V(t)\, \underline{z}^k$$

and thus, by (4.63), and the definition of z_-^* :

$$\lim_{t \to +\infty} d\,(V(t)\, z^o\,,\, [z_-^*, +\infty)) = 0\,. \tag{4.64}$$

From the above derivation it is clear that

$$0 \ll z_-^* \leq z_+^*$$

so that together (4.62) and (4.64) imply

$$\lim_{t \to +\infty} d\,(V(t)\, z^o\,,\, [z_-^*, z_+^*]) = 0\,. \tag{4.65}$$

Now properties (F4) and (F5) of f imply (Lemma A.35 in Appendix A) that the evolution operator $V(t)$, $t \geq 0$ is strictly concave .

The following main theorem holds (see Theorem A.36 in Appendix A).

Theorem 4.9. *If $\mu > 0$, then system (4.43) admits a unique nontrivial endemic state $z^* \gg 0$ which is GAS in $\dot{\mathcal{R}}$.*

In [18] an extension of the above analysis is carried out to the case in which the Lajmanovich-Yorke model exhibits seasonality in the parameters.

Actually, also this case can be reconsidered along the lines of Sect. 4.5 .

4.6.2. Macdonald' s model for the transmission of schistosomiasis in heterogeneous populations [22]

In Sect. 4.3.2 we have introduced Macdonald' s model for schistosomiasis in an homogeneous population as described by system (4.8).

In the linear case (4.9) it can be written as follows

$$(4.66)\qquad \begin{cases} \dfrac{dX}{dt} = \alpha\sigma Y - \gamma X \\ \dfrac{dY}{dt} = \beta X \left(1 - \dfrac{Y}{N}\right) - \delta Y \end{cases}$$

Here we remind that X denotes the current number of egg laying schistosomes in the human host population, while Y denotes the current number of infected snails in the environment (N is the constant total population of snails).

If individual and spatial heterogeneity within the community is taken into account, the following extension of Macdonald' s model proposed by Barbour [16] seems appropriate.

Suppose the community consists of M people living in an area where there are L ponds, and that person i ($i = 1, \ldots, M$) spends an amount of time $\lambda_{ij} \geq 0$ per day in pond j ($j = 1, \ldots, L$), where the snail density is ρ_j and the accessible water area A_j, $j = 1, \ldots, L$. Then the universal contact rates α and β are replaced by a set of rates

$$\alpha_{ij} = \alpha\lambda_{ij} \quad , \quad \beta_{ij} = \beta\lambda_{ij}, \qquad i = 1, \ldots, M \ ; \ j = 1, \ldots, L \ ,$$

and system (4.66) is replaced by the following family of equations for X_i , the number of female schistosomes within person i ($i = 1, \ldots, M$), and Y_j , the number of infected snails in pond j ($j = 1, \ldots, L$) :

$$(4.67)\qquad \begin{cases} \dfrac{d}{dt} X_i = \alpha \displaystyle\sum_{j=1}^{L} \dfrac{\lambda_{ij}}{A_j} Y_j - \gamma X_i \ , & i = 1, \ldots, M \\ \dfrac{d}{dt} Y_j = \beta \displaystyle\sum_{i=1}^{M} X_i \lambda_{ij} \left(1 - \dfrac{Y_j}{\rho_j A_j}\right) - \delta Y_j \ , & j = 1, \ldots, L \ . \end{cases}$$

Let us denote by $z := (X_1, \ldots, X_M, Y_1, \ldots, Y_L)^T$ and by

$$f := z \in \mathbb{R}_+^{M+L} \longrightarrow f(z) = (f_1(z), \dots, f_{M+L}(z))^T \in \mathbb{R}^{M+L}$$

such that

$$(4.68)\qquad \begin{aligned} f_i(z) &= \alpha \sum_{j=M+1}^{M+L} \frac{\tilde{\lambda}_{ij}}{\tilde{A}_j} z_j - \gamma z_i\,, && i = 1, \dots, M \\ f_j(z) &= \beta \sum_{i=1}^{M} z_i \tilde{\lambda}_{ij} \left(1 - \frac{z_j}{\tilde{\rho}_j \tilde{A}_j}\right) - \delta z_j\,, && j = M+1, \dots, M+L \end{aligned}$$

where

$$(4.69)\qquad \begin{aligned} \tilde{\lambda}_{ij} &= \lambda_{i,j-M} \\ \tilde{\rho}_j &= \rho_{j-M} \\ \tilde{A}_j &= A_{j-M} \end{aligned}$$

for $i = 1, \dots, M$ and $j = M+1, \dots, M+L$. Then system (4.67) can be rewritten in vector form

$$(4.70)\qquad \frac{dz}{dt} = f(z)\,, \qquad t > 0\,.$$

The Jacobi matrix of f at any point $z \in \mathbb{R}^{M+L}$ is the following

$$(4.71a)\qquad (J_f(z))_{ij} = \alpha \frac{\tilde{\lambda}_{ij}}{\tilde{A}_j}$$

$$(4.71b)\qquad (J_f(z))_{ji} = \beta \tilde{\lambda}_{ji} \left(1 - \frac{z_j}{\tilde{\rho}_j \tilde{A}_j}\right)$$

for $i = 1, \dots, M$, $j = M+1, \dots, M+L$, $i \neq j$;

$$(4.71c) \qquad (J_f(z))_{ii} = \begin{cases} -\gamma & i = 1, \dots, M \\ -\delta & i = M+1, \dots, M+L \end{cases}$$

$$(4.71d) \qquad (J_f(z))_{ij} = 0 \qquad \text{otherwise} .$$

If we assume that the contact matrix $\Lambda := (\lambda_{ij})_{i=1,\dots,M;j=1,\dots,L}$ is nonnegative and irreducible, then the Jacobi matrix $J_f(z)$ will be so for any $z \in \mathcal{R}$, defined as follows

$$(4.72) \qquad \mathcal{R} := \left\{ z \in \mathbb{R}_+^{M+L} \,\middle|\, z_j \leq \tilde{\rho}_j \tilde{A}_j \,, \quad j = M+1, \dots, M+L \right\} .$$

Further, we have the following usual properties for f :

(F1) $f(0) = 0$

(F2) f is quasimonotone nondecreasing in $\mathcal{R}$

(F3) for any $z \in \mathcal{R}$ there exists a $z^o \in \mathring{\mathcal{R}}$ such that $z \leq z^o$ and $f(z^o) \ll 0$.

The initial value problem associated with (4.28) is well posed for any choice of the initial condition in $\mathcal{R}$. Thanks to (F3) a bounded invariant rectangle can be defined for any solution subject to an initial condition in $\mathcal{R}$, and this implies that such a solution can be extended to the whole $\mathbb{R}_+$.

In particular it can be easily proven that

Lemma 4.10. *$\mathcal{R}$ is positively invariant for system (4.70). Further we have that*

$$(4.73) \qquad \tilde{\mathcal{R}} := \left\{ z \in \mathcal{R} \,\middle|\, z_i \leq \frac{\alpha}{\gamma} \sum_{j=1}^{L} \lambda_{ij} \, \rho_j \,, \quad i = 1, \dots, m \right\}$$

is positively invariant for system (4.70) and

$$(4.74) \qquad \textit{for any} \quad z^o \in \mathcal{R} : \quad \lim_{t \to +\infty} d\,[V(t)\, z^o \,, \tilde{\mathcal{R}}] = 0 \,.$$

As usual we shall denote the evolution operator of system (4.70) by $\{V(t), t \geq 0\}$ so that the solution of the system subject to the initial condition $z^o \in \mathcal{R}$ will be given by $z(t) = V(t)\, z^o, \quad t \geq 0$.

Due to the quasimonotonicity of f in $\mathcal{R}$ (condition (F2)) we have that for any $t \geq 0$, $V(t)$ is a monotone nondecreasing operator, in the partially ordered space $\mathbb{R}_+^{M+L}$ $(\cap \mathcal{R})$:

$$\text{(4.75)} \quad \text{for any } z^1, z^2 \in \mathcal{R},\ z^1 \leq z^2 \implies V(t)\, z^1 \leq V(t)\, z^2 \text{ for any } t \geq 0.$$

The Jacobi matrix $J_f(z)$ is monotone nonincreasing with respect to z in $\overset{\circ}{\mathcal{R}}$:

$$\text{(4.76)} \qquad \text{for any} \quad z', z'' \in \overset{\circ}{\mathcal{R}}, \quad z' < z'' :$$
$$(J_f(z'))_{ij} \geq (J_f(z''))_{ij}\,, \qquad i,j = 1, \ldots, M+L.$$

This implies condition (F4) (see Appendix A, Section A.4.2) so that by Lemma A.34 in Appendix A the strong monotonicity of the evolution operators follows

$$\text{(4.77)} \quad \text{for any } z^1, z^2 \in \overset{\circ}{\mathcal{R}},\ z^1 \leq z^2,\ z^1 \neq z^2 : \quad V(t)\, z^1 \ll V(t)\, z^2,\ t > 0.$$

From the definition (4.68) we also obtain the sublinearity of f

$$\text{(F5)} \text{ for any } \ z \in \overset{\circ}{\mathcal{R}}, \ \text{ for any } \tau \in (0,1) : \quad f(\tau z) > \tau f(z).$$

As a consequence

$$\text{(4.78)} \qquad \text{for any} \quad z \in \overset{\circ}{\mathcal{R}} : \qquad f(z) \leq J_f(0)\, z$$

where $J_f(0)$ is the Jacobi matrix of f at zero :

$$\text{(4.79}a\text{)} \qquad (J_f(0))_{ij} = \alpha \frac{\tilde{\lambda}_{ij}}{\tilde{A}_j}\,,$$

$$\text{(4.79}b\text{)} \qquad (J_f(0))_{ji} = \beta \tilde{\lambda}_{ji}\,,$$

for $i = 1, \ldots, M\,;\ j = M+1, \ldots, M+L\ ;$

$$(J_f(0))_{ii} = \begin{cases} -\gamma\,, & i = 1, \ldots, M \\ -\delta\,, & i = M+1, \ldots, M+L \end{cases} \;; \tag{4.79c}$$

$$(J_f(0))_{ij} = 0 \qquad \text{otherwise}\,. \tag{4.79d}$$

The following lemma holds.

Lemma 4.11. *If Λ is irreducible, then*

$$z^o \in \dot{\mathcal{R}} \quad \Longrightarrow \quad V(t)\, z^o \in \mathring{\mathcal{R}} \tag{4.80}$$

i.e. the evolution operator $V(t)$ of system (4.70) is strongly positive. Moreover

$$z^o \in \dot{\tilde{\mathcal{R}}} \quad \Longrightarrow \quad V(t)\, z^o \in \mathring{\tilde{\mathcal{R}}}\,. \tag{4.81}$$

The biological meaning of Lemma 4.11 is that when Λ is irreducible, no pond can have all snails uninfected for a positive time interval; furthermore no individual can be free of schistosomes for a positive time interval, unless every individual and all snails are free of the disease.

In fact system (4.70) always admits the trivial equilibrium solution $\mathbf{0} := (0, \ldots, 0)^T$.

As a consequence of the fact that $J_f(0)$ is a quasimonotone irreducible matrix, a real dominant eigenvalue μ will exist equal to the spectral radius of the matrix $J_f(0)$. The stability properties of the trivial solution are based on the sign of μ.

Theorem 4.12. *Let $\mu \in \mathbb{R}$ be the dominant real eigenvalue of $J_f(0)$.*

a) If $\mu \leq 0$, then the trivial equilibrium is GAS in $\mathcal{R}$ for system (4.70).
b) If $\mu > 0$, then the trivial equilibrium is unstable for system (4.70). Further there exists a $k > 0$ such that

$$\textit{for any} \quad z^o \in \dot{\mathcal{R}}: \qquad \liminf_{t \to +\infty} |V(t)\, z^o| \geq k\,.$$

Proof. This theorem is a direct consequence of Theorem B.27 and Theorem B.28 in Appendix B.

Under assumption b) of Theorem 4.12 we may proceed as in Sect. 4.6.1 (see Eqn. (4.58) and following) to look for nontrivial endemic solutions, i.e. of nontrivial solutions of system

$$f(z) = 0 \,. \tag{4.82}$$

As a consequence of (4.74) in Lemma 4.10 no equilibrium solution of system (4.70) can stay out of $\tilde{\mathcal{R}}$, so that we shall limit our research of nontrivial solutions of system (4.82) to $\tilde{\mathcal{R}}$.

As in Sect. 4.6.1 we can build up a sequence of nontrivial subsolutions of system (4.82), i.e. a sequence $\left(\underline{z}^k\right)_{k\in\mathbb{N}} \in \overset{\circ}{\tilde{\mathcal{R}}}$ such that

$$\text{for any} \quad k \in \mathbb{N} : \quad f\left(\underline{z}^k\right) \geq 0\,, \tag{4.83}$$

with the further property

$$\text{for any } z^o \in \overset{\circ}{\tilde{\mathcal{R}}} \text{ a } k \in \mathbb{N} \text{ exists such that } \; 0 \ll \underline{z}^k \leq z^o\,. \tag{4.84}$$

Further,

$$\overline{z} := \left(\frac{\alpha}{\gamma}\sum_{j=1}^{L}\lambda_{1j}\rho_j, \ldots, \frac{\alpha}{\gamma}\sum_{j=1}^{L}\lambda_{Mj}\rho_j, \rho_1 A_1, \ldots, \rho_L A_L\right)^T,$$

is a supersolution of system (4.82); in fact

$$f(\overline{z}) \leq 0\,. \tag{4.85}$$

For any choice of $z^o \in \dot{\tilde{\mathcal{R}}}$ it is possible then to identify a suitable subsolution $\underline{z} \in \overset{\circ}{\tilde{\mathcal{R}}}$ and a supersolution $\overline{z}$ such

$$\underline{z} \leq z^o \leq \overline{z}\,.$$

As a consequence of the comparison theorem (Appendix A, Section A.4.2)

$$\text{(4.86)} \qquad \text{for any} \quad t > 0 : \qquad V(t)\,\underline{z} \le V(t)\,z^o \le V(t)\,\overline{z}$$

where, because of Lemma B.29 and Lemma B.30 in the Appendix B, $V(t)\,\underline{z}$ is monotone nondecreasing in $t > 0$; and $V(t)\,\overline{z}$ is monotone nonincreasing in $t > 0$.

Further two nontrivial equilibrium solutions $z^*_-, z^*_+ \in \overset{\circ}{\tilde{\mathcal{R}}}$ exist such that

$$\text{(4.87)} \qquad \lim_{t \to +\infty} d\left(V(t)\,z^o, [z^*_-, z^*_+]\right) = 0\,.$$

Now, properties (F4) and (F5) of f imply (Lemma A.35 in Appendix A) that the evolution operator $V(t)$, $t \ge 0$ is strongly concave; this fact, together with the strong positivity and the monotonicity, implies the following (see Theorem A.36 in the Appendix A)

Theorem 4.13. *If $\mu > 0$ is the dominant real eigenvalue of $J_f(0)$, then system (4.70) admits a unique nontrivial endemic state $z^* \in \overset{\circ}{\tilde{\mathcal{R}}}$ which is GAS in $\dot{\mathcal{R}}$.*

As far as the existence and uniqueness of the nontrivial endemic state z^* we wish to derive the equivalent condition (2) of Appendix 2 in [22].

From the definition (4.68) of f we derive that the components of a nontrivial equilibrium solution of system (4.70) must satisfy

$$\text{(4.88)} \qquad \begin{cases} \alpha \displaystyle\sum_{k=1}^{L} \frac{\lambda_{ik}}{A_k} Y_k = \gamma X_i \,, & i = 1, \dots, M \\ \beta \displaystyle\sum_{i=1}^{M} X_i \lambda_{ij} \left(1 - \frac{Y_j}{\rho_j A_j}\right) = \delta Y_j \,, & j = 1, \dots, L \end{cases}$$

where we have denoted again by $X_1, \dots, X_M$ the first M components of the state vector z , and by $Y_1, \dots, Y_L$ the other L components of z .

By solving the first of (4.88) with respect to X_i, $i = 1, \dots, M$, and substituting into the second we get

$$\text{(4.89)} \qquad Y_j = \frac{\dfrac{\beta\alpha}{\delta\gamma} \displaystyle\sum_{i=1}^{M} \sum_{k=1}^{L} \frac{\lambda_{ik}}{A_k} \lambda_{ij} Y_k}{1 + \dfrac{\beta\alpha}{\delta\gamma\rho_j A_j} \displaystyle\sum_{i=1}^{M} \sum_{k=1}^{L} \frac{\lambda_{ik}}{A_k} \lambda_{ij} Y_k}$$

that can be seen, in vector form, as

$$Y = F(Y) \tag{4.90}$$

by defining the components of $F(Y)$ as the right hand side of (4.89) .

The map F is a continuous monotone nondecreasing, strictly sublinear, bounded function which maps the interval $I^L := [0, \rho_1 A_1] \times \cdots \times [0, \rho_L A_L]$ into itself. By strictly sublinear we mean the following : for any $Y \in I^L$, for any $\tau \in (0,1)$ there exists an $\epsilon > 0$ such that $F(\tau Y) \geq (1+\epsilon)\tau F(Y)$ [137].

Clearly $F(0) = 0$. Furthermore the Jacobi matrix of F at 0 is given by

$$(J_F(0))_{jk} = \frac{\alpha\beta}{\gamma\delta A_j} \sum_{i=1}^{M} \lambda_{ij}\lambda_{ik} \tag{4.91}$$

which is irreducible if Λ is so .

According to a well known result [137, 142, 114] we can state the following

Theorem 4.14. *F does not have nontrivial fixed points on the boundary of I^L . Moreover F has a positive fixed point iff the spectral radius $\rho(J_F(0)) > 1$. If there is a positive fixed point, then it is unique.*

Remark. It can be easily seen from (4.91) that the condition $\rho(J_F(0)) > 1$ reduces to the threshold condition in Sect. 4.3.2 for $L = M = 1$.

Thus $\rho(J_F(0))$ is the natural generalization of the threshold parameter $R := \dfrac{\alpha\beta\sigma}{\gamma\delta}$ (for a discussion about the role of R , the basic reproduction ratio, see [9] and [79]).

We may further remark that when we assume that the elements of Λ are standardized by the requirement that $\sum_{i,j} \lambda_{ij} = M > 0$, $\rho(J_F(0))$ is minimized among all possible choices of λ_{ij} by the homogeneous mixing choice

$$\lambda_{ij} = \frac{A_j}{A}$$

in which case

$$\rho(J_F(0)) = \frac{\alpha\beta\sigma}{\gamma\delta}$$

where now $\sigma = M/A$ (for the proof of this statement we refer to [22]).

For the biological discussion, and for the problem of parameter estimation we also refer to [22].

5. Spatial heterogeneity

5.1. Introduction

The role of spatial heterogeneity and dispersal in population dynamics has been the subject of much research. We shall refer here to the fundamental literature on the subject by quoting [12, 17, 86, 87, 108, 131, 148, 152, 170, 177, 201, 204].

In the theory of propagation of infectious diseases the motivation of the study of the effects of spatial diffusion is mainly due to the large scale impact of an epidemic phenomenon; this is discussed in detail in [69].

We will choose here as a good starting point the pioneer work by D.G. Kendall who modified the basic Kermack-McKendrick SIR model to include the effects of spatial heterogeneity in an epidemic system [130, 131].

Kendall 's work has motivated a lot of research in the theory of epidemics with spatial diffusion. In particular two classes of problems arise according to the size of the spatial domain or habitat.

If the habitat is unbounded, travelling waves are of interest [131, 121, 15, 78, 207]. A nice introduction to the subject can be found in [171].

If, on the other hand, the habitat is a bounded spatial domain then problems of existence of nontrivial endemic states (possibly with spatial patterns) are of interest [22, 45, 53, 54]. This will be the subject of our presentation.

Usually reaction diffusion systems (see e.g. [86]) are seen as an extension of compartmental systems in which each compartment, representing a different species, is allowed to invade a spatial domain $\Omega \subset \mathbb{R}^m$ with a space dependent density. Densities interact among themselves according to the same mathematical laws which were used for the space independent case, but are subject individually to a spatial diffusion mechanism usually committed to the Laplace operator, which simulates random walk or Brownian motion of the interacting species [16, 86].

Typically then a system of n interacting species each of them having a spatial density

$$\{u_i(x;t), \quad x \in \Omega\}, \quad i = 1, \ldots, n \quad \text{at time} \quad t \geq 0$$

is described by the following system of semilinear parabolic equations:

$$\frac{\partial u}{\partial t} = D\Delta u(x;t) + f(u(x;t)) \tag{5.1}$$

in $\Omega \times \mathbb{R}_+$, subject to suitable boundary conditions.
Here $u = (u_1, \ldots, u_n)^T$; $D = diag(d_1, \ldots, d_n)$, and $f(z)$, $z \in \mathbb{R}^n$ is the interaction law among the species via their densities.

A derivation and discussion of equation (5.1) can be found in [86] (see also [177]).

In equation (5.1) $f(z)$, $z \in \mathbb{R}^n$, usually is the same interaction function of the classical compartmental approach. Thus for an SIR model with vital dynamics system (5.1) is written as follows

$$\left\{\begin{aligned} \frac{\partial s}{\partial t}(x;t) &= d_1 \Delta s(x;t) - k\, i(x;t)\, s(x;t) + \mu - \mu\, s(x;t) \\ \frac{\partial i}{\partial t}(x;t) &= d_2 \Delta i(x;t) + k\, i(x;t)\, s(x;t) - (\mu + \gamma)\, i(x;t) \\ \frac{\partial r}{\partial t}(x;t) &= d_3 \Delta r(x;t) + \gamma\, i(x;t) - \mu\, r(x;t) \end{aligned}\right. \tag{5.2}$$

where now s, i, r are the corresponding spatial densities of S, I, R :

$$S(t) = \int_\Omega s(x;t)\,dx; \quad I(t) = \int_\Omega i(x;t)\,dx : \quad R(t) = \int_\Omega r(x;t)\,dx$$

Thus the infection process is represented by a "local" interaction of two densities $s(x;t)$ and $i(x;t)$ at point $x \in \Omega$ and time $t \in \mathbb{R}_+$ via the following "law of mass action".

$$k\, i(x;t)\, s(x;t) \tag{5.3}$$

In Kendall 's model [130, 131] (5.3) is substituted by an "integral" interaction between the infectives and the susceptibles; the force of infection acting at point $x \in \Omega$ and time $t \in \mathbb{R}_+$ is given by

$$g(i(\cdot;t))(x) = \int_\Omega k(x,x')\, i(x';t)\,dx' \tag{5.4}$$

where $k(x,x')$ describes the influence, by any reason, of the infectives located at any point $x' \in \Omega$ on the susceptibles located at point $x \in \Omega$.

As a consequence the infection process is described now by

$$g(i(\cdot;t))(x)\, s(x;t) \ . \tag{5.5}$$

Clearly we reobtain (5.3) in the limiting case $k(x, x') = \delta(x - x')$ (the Dirac function).

As one can see (5.5) better fits the philosophy introduced in Section 1 to allow $g(I)$, the force of infection due to the population of infectives, and acting on the susceptible population, to have a general form which case by case takes into account the possible mechanisms of transmission of the disease.

For problem (5.2) we refer to the literature [15, 45, 78, 207]. In Sect. 5.6 an extension of the Lyapunov methods discussed in Sect. 2 will be presented.

By now we shall limit our analysis to quasimonotone reaction-diffusion systems, and, as an application, to the following system which extends to the space heterogeneous case our "study model" (4.18) for positive feedback systems [57],

$$(5.6) \qquad \begin{cases} \dfrac{\partial u_1}{\partial t}(x;t) = d_1 \Delta u_1(x;t) - a_{11} u_1(x;t) + a_{12} u_2(x;t) \\ \dfrac{\partial u_2}{\partial t}(x;t) = d_2 \Delta u_2(x;t) - a_{22} u_2(x;t) + g(u_1(x;t)) \end{cases}$$

in $\Omega \times \mathbb{R}_+$, subject to suitable boundary and initial conditions.

By analyzing system (5.6) and subsequent modifications we shall show how robust quasimonotone systems are structurally.

This is particularly true when g is concave. In this case in fact the qualitative behavior of the system does not change with respect to various modifications of the infection process.

The approach is based on [161] and [57].

When monotonicity properties do not hold for (5.1), extensions of Lyapunov methods and of the LaSalle Invariance Principle to dynamical systems in Banach spaces (Appendix B) may still be taken into account to analyze the asymptotic behavior of the system.

In Section 5.6 this approach is presented by reconsidering the general model of Section 2.3 subject to spatial diffusion.

This is essentially based on [187] and [49].

5.2. Quasimonotone systems

As we did in the previous chapters our aim here is not to give a complete review of all the literature about quasimonotone systems and their applications. On the other hand we shall present the basic theory of such systems, and apply it to the simpler nonlinear epidemic model (4.18) as a case study. Again we shall distinguish two cases, according to the concavity assumptions on g, (v) or (v') of Section 4.4. The possibility of seasonal variation of g will be analyzed too.

A brief account of the mathematical theory is reported in Appendix B, Section B.2.1 .

We shall consider the following reaction-diffusion system

$$\frac{\partial}{\partial t}u(x;t) = D\Delta u(x;t) + f(u(x;t)) \ , \tag{5.7}$$

for $x \in \Omega$, and $t > 0$.

Here $u := (u_1, u_2)^T$ is a vector of two real functions defined in $\Omega \in \mathbb{R}^m$.

We shall suppose that Ω is an open bounded subset of $\mathbb{R}^m$ (we shall limit ourselves to the case $m = 1, 2, 3$), with a sufficiently smooth boundary $\partial\Omega$.

Classical boundary conditions will be imposed on (5.7).

$$B_i u_i(x;t) := \frac{\partial}{\partial \nu}u_i(x;t) + \alpha_i(x)u_i(x;t) = 0 \ , \quad i = 1, 2 \tag{5.7b}$$

for $x \in \partial\Omega$, and $t > 0$; here $\frac{\partial}{\partial \nu}$ denotes the outward normal derivative, and $\alpha_i(x) \geq 0$, $i = 1, 2$ is chosen to be a sufficiently smooth function of $x \in \partial\Omega$. For $\alpha_i = 0$ we get homogeneous Neumann boundary conditions; for $\alpha_i = +\infty$ we get homogeneous Dirichlet boundary conditions (unless explicitly stated, we shall ignore this case), $i = 1, 2$. D is the diagonal matrix of the diffusion coefficients, $D = diag(d_1, d_2)$ and we shall assume at first that both $d_i > 0, i = 1, 2$.

Finally we choose $f : \mathbb{K} \to \mathbb{R}^2$ as follows (see Eqn. (4.19)):

$$\begin{aligned} z = (z_1, z_2)^T \in \mathbb{K} \to f(z) &= \begin{pmatrix} f_1(z_1, z_2) \\ \\ f_2(z_1, z_2) \end{pmatrix} \\ &= \begin{pmatrix} -a_{11}z_1 + a_{12}z_2 \\ \\ g(z_1) - a_{22}z_2 \end{pmatrix} \in \mathbb{R}^2 \end{aligned} \tag{5.8}$$

with g satisfying conditions (i)-(iv) in Section 4.4, and a_{ij} positive real numbers, $i, j = 1, 2$.

As a consequence, f satisfies the usual conditions

(F1) $f(0) = 0$

(F2) f is quasimonotone increasing in $\mathbb{K}$;

(F3) for any $z \in \mathbb{K}$, a $z_o \in \mathring{\mathbb{K}}$, $z \leq z_o$ exists such that $f(z_o) \ll 0$.

System (5.7), (5.7b) has to be supplemented by an initial condition

$$u(x;0) = u^o(x), \qquad x \in \Omega \ . \tag{5.9}$$

We shall assume here that all conditions are satisfied for the existence and uniqueness of a classical solution u of problem (5.7), (5.7b), (5.9) (see Appendix B), which can be extended to all $t \in \mathbb{R}_+$ because of condition (F3).

By classical solution we mean that

$$u \in C^{2,1}\left(\Omega \times (0,\infty), \mathbb{R}^2\right) \cap C^{1,0}\left(\overline{\Omega} \times (0,\infty), \mathbb{R}^2\right).$$

We shall denote by $U(t)u^o$ the unique solution of problem (5.7), (5.7b) subject to the initial condition (5.9).

It can be shown (Appendix B, Section B.2.1) that the family $\{U(t),\ t \in \mathbb{R}_+\}$ of evolution operators is a strongly continuous semigroup, i.e. it satisfies the following properties

(5.10a) $U(0) = I$

(5.10b) for any $s, t \geq 0 : U(t+s) = U(t)U(s)$

(5.10c) for any $t \geq 0 : U(t)0 = 0$

(5.10d) for any $t \geq 0$, the mapping

$$u^o \in X_+ \longrightarrow U(t)u^o \ \in X$$

is continuous, uniformly in $t \in [t_1, t_2] \subset \mathbb{R}_+$

(5.10e) for any $u^o \in X$, the mapping

$$t \in [0, +\infty) \longrightarrow U(t)u^o$$

is continuous.

Due to the assumptions on g, and the quasimonotonicity of f in $\mathbb{K}$, we have that $\{U(t),\ t \in \mathbb{R}_+\}$ is a family of monotone operators in the (partially) ordered Banach space $X := C(\overline{\Omega}) \times C(\overline{\Omega})$ (see Appendix B, Section B.1.1), i.e.

$$\text{(5.11)} \quad \text{for any } u^o, v^o \in X_+, u^o \le v^o \Longrightarrow U(t)u^o \le U(t)v^o \text{ for any } t \in \mathbb{R}_+.$$

In particular we have the positivity of $\{U(t),\ t \in \mathbb{R}_+\}$; i.e.

$$\text{(5.11')} \qquad \text{for any } \ u^o \in X_+ \ : \ U(t)u^o \in X_+ \ , \qquad t \in \mathbb{R}_+$$

In this section, we shall assume further that g satisfies condition (v) of Section 4.4

(v) $0 < g'(0+) < +\infty$, and for any $z > 0$: $g''(z) < 0$

As a consequence the Jacobi matrix of f

$$\text{(5.12)} \qquad J_f(z) = \begin{pmatrix} -a_{11} & a_{12} \\ g'(z_1) & -a_{22} \end{pmatrix}$$

is irreducible for any $z \ge 0$.

In particular $J_f(0+)$ is quasimonotone increasing and irreducible.

This implies (Lemma B.20 in Appendix B) that $U(t)$ is strongly positive for $t > 0$; i.e.

$$\text{(5.12')} \qquad \text{for any } \ u^o \in X_+ \ : \ u^o \ne 0 \implies U(t)u^o \gg 0, \quad t > 0 \ .$$

If we now take into account that g is also concave, then $J_f(z)$ is nondecreasing in z; i.e.

$$0 < z' < z'' \quad \Longrightarrow \quad (J_f(z'))_{ij} \ge (J_f(z''))_{ij} \quad \text{for} \quad i,j = 1,\dots,m.$$

This property implies the following

(F4) for any $R > 0$ an irreducible quasimonotone increasing 2x2 matrix C_R exists such that

for any $\quad z', z'' \in \mathbb{K}, \quad 0 \leq z' \leq z''; \quad |z'|, |z''| \leq R$:

$$f(z'') - f(z') \geq C_R(z'' - z') \ .$$

In fact for $\quad 0 \leq z' \leq z'' \quad$ we have

$$f(z'') - f(z') = \int_0^1 J_f(tz'' + (1-t)z')(z'' - z')\, dt \ .$$

Hence, if $\zeta \in \mathbb{K}$ is such that for any $z \in \mathbb{K}, |z| \leq R: \quad z \leq \zeta$, from the monotonicity of J_f we get

$$f(z'') - f(z') \geq \int_0^1 J_f(\zeta)(z'' - z')\, dt \ ,$$

from which (F4) follows. Lemma B.21 in Appendix B implies the strong monotonicity of the evolution operator $\{U(t), t \in \mathbb{R}_+\}$:

(5.13) for any $\quad u^o, v^o \in X_+, u^o \leq v^o, u^o \neq v^o : U(t)u^o \ll U(t)v^o, \quad t > 0$.

From the strict concavity of g it also follows that

(5.14) for any $\quad {z'}_1, {z''}_1 \in \mathbb{R}_+, \quad {z'}_1 \neq {z''}_1 : \quad$ for any $\quad \tau \in (0,1)$:

$$g(\tau {z'}_1 + (1-\tau){z''}_1) > \tau g({z'}_1) + (1-\tau) g({z''}_1) \ ;$$

and in particular for ${z''}_1 = 0$:

(5.14′) for any $\quad z_1 \in \mathbb{R}_+ - \{0\}; \quad$ for any $\quad \tau \in (0,1) : \quad g(\tau z_1) > \tau g(z_1)$

This implies sublinearity:

(F5) for any $\quad z \in \mathbb{K}, z \gg 0; \quad$ for any $\quad \tau \in (0,1) : f(\tau z) > \tau f(z)$.

As a consequence

$$\text{for any} \quad z \in \mathbb{K}, z \gg 0; \quad \text{for any} \quad \tau \in (0,1) : f(z) < \frac{1}{\tau} f(\tau z)$$

and for $\tau \to 0^+$:

$$\text{for any} \quad z \in \mathbb{K} : \quad f(z) \leq Bz \tag{5.15}$$

where

$$B = \begin{pmatrix} -a_{11} & a_{12} \\ g'(0+) & -a_{22} \end{pmatrix} \tag{5.16}$$

We may also note, as we did in the analysis of the periodic ODE case, (Lemma 4.4) that, due to the strict concavity of g (assumption (v)), we may also state that, for any $z \in \overset{\circ}{\mathbb{K}} : zJ_z(z) - f(z)$ has nonpositive components and one of them is negative.

In the sequel we shall denote by

$$\alpha_m := \min_{x \in \partial\Omega} \min \{\alpha_1(x), \alpha_2(x)\} \tag{5.17a}$$

$$\alpha_M := \max_{x \in \Omega} \max \{\alpha_1(x), \alpha_2(x)\} \tag{5.17b}$$

Correspondingly we shall denote by λ_m, respectively λ_M, the corresponding eigenvalues of the boundary value problem (B.14) in Appendix B, Section B.2.1.1 . Let ϕ_m and ϕ_M be the corresponding eigenfunctions.

The following theorems hold.

Theorem 5.1. *If*

$$\theta_m := \frac{a_{12} g'(0+)}{(a_{11} + d_1 \lambda_m)(a_{22} + d_2 \lambda_m)} < 1 \tag{5.18}$$

then the trivial solution is GAS in X_+ for system (5.7), (5.7b)

Proof. Due to (5.15) we have that

$$\text{for any} \quad z \in \mathbb{K}: \quad f(z) \leq Bz \ .$$

By the comparison Theorem B.19 in the Appendix B with both $\alpha_i(x) = \alpha_m, \quad i = 1, 2,$ we can state that

$$\text{for any} \quad v^o \in X_+ : 0 \leq U(t)v^o \leq T_m v^o \quad \text{in} \quad (0, +\infty) \ . \tag{5.19}$$

where T_m is the linear evolution operator of the linear system

$$\frac{\partial}{\partial t} v(x;t) = D\Delta v(x;t) + Bv(x;t) \tag{5.20}$$

for $x \in \Omega$, and $t > 0$, with boundary conditions

$$\frac{\partial}{\partial \nu} v(x;t) + \alpha_m v(x;t) = 0 \tag{5.20b}$$

for $x \in \partial\Omega$, and $t > 0$.

For $\theta_m < 1$, it easily seen that both eigenvalues of the matrix $(-\lambda_m D + B)$ are negative. The theorem follows from Theorem B.27 in Appendix B.

Theorem 5.2. *If*

$$\theta_M := \frac{a_{12} g'(0+)}{(a_{11} + d_1 \lambda_M)(a_{22} + d_2 \lambda_M)} > 1 \tag{5.21}$$

then the trivial solution is unstable for system (5.7), (5.7b).

Proof. Let B be defined as in (5.16) and observe that, due to (5.21), the largest eigenvalue of $(-\lambda_M D + B)$ is positive.

Now, choose $\epsilon \in \mathbb{R}, \ 0 < \epsilon < g'(0+)$, such that the largest eigenvalue μ_ϵ of $(\lambda_M D + B_\epsilon)$ is still positive, where

$$B_\epsilon := \begin{pmatrix} -a_{11} & a_{12} \\ g'(0+) - \epsilon & -a_{22} \end{pmatrix} .$$

Due to the basic assumptions on g, a $\delta > 0$ exists such that

$$\text{(5.22)} \qquad \text{for any} \quad z \in \mathbb{K}, \quad |z| \leq \delta : \quad f(z) \geq B_\epsilon z \, .$$

Now, the matrix $(-\lambda_M D + B)$ is a quasimonotone increasing irreducible matrix. By the Perron-Frobenious theorem (Appendix A, Section A.4.2) we can state that, corresponding to its largest eigenvalue $\mu_\epsilon > 0$, it is possible to associate an eigenvector $\eta \in \mathbb{K}, \eta \gg 0$.

We may apply (5.13) to $T_\epsilon(t)$, the evolution operator of the linear system

$$\text{(5.23)} \qquad \frac{\partial}{\partial t} v(x;t) = D\Delta v(x;t) + B_\epsilon v(x;t)$$

for $x \in \Omega$, and $t > 0$, with boundary conditions

$$\text{(5.23}b\text{)} \qquad \frac{\partial}{\partial \nu} v(x;t) + \alpha_M v(x;t) = 0$$

for $x \in \partial\Omega$, and $t > 0$. We get in particular that

$$\text{(5.24)} \quad \text{for any} \quad t > 0, \quad \text{for any} \quad v^o \in X_+, v^o \neq 0, T_\epsilon(t)v^o \gg 0, \quad \text{in} \quad \overline{\Omega}.$$

It is well known that, whenever $\phi, \psi, : \overline{\Omega} \to \mathbb{R}$ are such that $\phi, \psi > 0$ in $\overline{\Omega}$, then a $\gamma > 0$ exists such that for any $x \in \overline{\Omega} : \gamma\phi(x) \leq \psi(x)$.

If we apply this fact to $T_\epsilon(t)v^o$ and $\eta\, \phi_M$ we shall have

$$\gamma\eta\, \phi_M \leq T_\epsilon(t)v^o$$

for a suitable choice of $\gamma > 0$.

As a consequence, for any $s \geq 0$ and $t > 0$:

$$T_\epsilon(t+s)v^o = T_\epsilon(s)T_\epsilon(t)v^o \geq T_\epsilon(s)\gamma\eta\, \phi_M = w_{\gamma\eta}(s)\, \phi_M$$

where, for the particular choice of η we have

$$w_{\gamma\eta}(s) = e^{\mu_\epsilon s}\gamma\eta$$

from which the theorem follows, being $\mu_\epsilon > 0$.

Observe that diffusion has for quasimonotone increasing systems a stabilizing effect of the trivial solution. In fact, if we recall the definition of

$$\theta := \frac{a_{12}\, g'(0+)}{a_{11}a_{22}} \tag{4.20}$$

we clearly have that

$$\theta_M \leq \theta_m \leq \theta \tag{5.25}$$

the equality sign holding if and only if $\alpha_i(x) = 0$ on $\partial\Omega$ i.e. for pure Neumann boundary conditions.

So that θ_m can be less than one even if θ is not such. Viceversa if $\theta < 1$ surely $\theta_m < 1$, so that the ODE system can be seen as a "pessimistic" model of the epidemic system with respect to the corresponding model with diffusion. On the other hand we have that if $\theta_M > 1$, also $\theta > 1$ so that, by Proposition 4.1, the ODE system (4.18) admits a GAS nontrivial endemic state $z^* \in \overset{\circ}{\mathbb{K}}$.

We shall prove now that this implies existence and GAS of a nontrivial endemic state (with a space structure in general) for the reaction-diffusion system (5.7), (5.7b).

By this we mean a function $\phi(x)$ which is a classical solution of the nonlinear elliptic problem

$$\begin{cases} D\Delta\phi(x) + f(\phi(x)) = 0, & \text{in } \Omega \\ B_i\phi_i(x) = 0, \quad i = 1, 2, & \text{on } \partial\Omega \end{cases} \tag{5.26}$$

In Appendix B, Section B.2.1.3, the notions of lower solution and of upper solution of system (5.26) are given. Comparison theorems are also reported there to prove existence and stability of solutions of (5.26) based on the existence of lower and upper solutions.

Here we shall make use of these techniques to solve our problem.

Proposition 5.3. *If $\theta_M > 1$, then the nontrivial equilibrium solution $z^* \in \overset{\circ}{\mathbb{K}}$ of the ODE system (4.18) is an upper solution of the nonlinear elliptic problem (5.26).*

Proof. It is a direct consequence of the definitions.

Observe further that when $\theta_M > 0$ then the ODE system

$$\frac{d}{dt}z(t) = -\lambda_M D z(t) + f(z(t)), \quad t > 0 \tag{5.27}$$

admits a unique nontrivial equilibrium solution z_M^* which is GAS in $\overset{\circ}{\mathbb{K}}$. This can be shown as in Theorem 4.2.

We shall make use of this fact to prove the following proposition.

Proposition 5.4. *If $\theta_M > 1$ then for any $\gamma > 0$ such that $\|\gamma\phi_M\| \leq 1$, the vector function*

$$\underline{\phi}_\gamma(x) := \gamma\,\phi_M(x)\,z_M^*, \quad \text{in } \overline{\Omega} \tag{5.28}$$

is a lower solution of the nonlinear elliptic system (5.26).

Proof. Thanks to property (F5) of the function f, we have

$$0 = -\lambda_M D\left(\gamma\phi_M z_M^*\right) + \gamma\phi_M f(z_M^*) < D\Delta\underline{\phi}_\gamma + f\left(\underline{\phi}_\gamma\right)$$

which proves the result by definition.

Once we have got a lower solution $\underline{\phi}_\gamma$ and an upper solution $\overline{\phi}$, under the above assumptions, we can then state that (Appendix B, Section B.2.1.3).

(a) $\left\{U(t)\underline{\phi}_\gamma; t \in \mathbb{R}_+\right\}$ is monotone nondecreasing in X_+;

(b) $\left\{U(t)\overline{\phi}; t \in \mathbb{R}_+\right\}$ is monotone nonincreasing in X_+;

(c) a minimal solution $\phi_-(x), x \in \overline{\Omega}$ and a maximal solution $\phi_+, x \in \overline{\Omega}$ exists for system (5.26) such that $0 \ll \phi_- \leq \phi_+$

and

(d) $\lim_{t\to+\infty} d\left(U(t)\underline{\phi}_\gamma, [\phi_-, +\infty)\right) = 0$

$\lim_{t\to+\infty} d\left(U(t)\overline{\phi}, [0, \phi_+]\right) = 0$

Remark. In the above a major role is played by the fact that f is a quasi-monotone function in $\mathbb{K}$, so that comparison theorems hold for our reaction diffusion system. It is worth noting that the technique of lower and upper solutions extends to parabolic systems with general classical boundary conditions the technique of contracting rectangles already used for the ODE case.

Clearly if the boundary conditions (5.7b) are pure Neumann boundary conditions then the two techniques are essentially equivalent, since we can always choose both lower and upper conditions to be space homogeneous [143].

At this point the problem of uniqueness and GAS stability of a nontrivial solution of system (5.26) is left open.

Actually under conditions (F4) and (F5) on f, by Lemma A.35 in Appendix A it can be shown that the evolution operator $\{U(t), t \in \mathbb{R}_+\}$ is strongly positive, monotone, and concave. As a consequence [139], it cannot have more than one fixed point. Since all solutions of system (5.26) are equilibrium solutions of system (5.7), (5.7b), i.e. fixed points for $\{U(t), t \in \mathbb{R}_+\}$, it must be

$$\phi_- = \phi_+ \ .$$

Let us denote by $\phi \in X_+, \quad 0 \ll \phi$ this unique solution of system (5.26).

As a consequence of (a), (b) and (d) we may state that for any lower solution $\underline{\phi}$ of system (5.26), $\{U(t)\underline{\phi},\ t \in \mathbb{R}_+\}$ converges monotonically (nondecreasing) to ϕ, while for any upper solution $\overline{\phi}$ of system (5.26), $\{U(t)\overline{\phi}, t \in \mathbb{R}_+\}$ converges monotonically (nonincreasing) to ϕ.

As a consequence we can finally prove the following main theorem.

Theorem 5.5. *If $\theta_M > 1$ system (5.7), (5.7b) admits a unique nontrivial endemic state $\phi(x), x \in \overline{\Omega}, \phi \gg 0$, which is GAS in $X_+ - \{0\}$.*

Proof. If $u^o \in X_+, u^o \neq 0$, due to property (F3) of f we can always choose a $z^o \in X_+, z^o \gg 0$, such that

$$0 \leq u^o \leq z^o \quad \text{and} \quad f(z^o) \leq 0 \ .$$

As a consequence z^o is an upper solution of system (5.26) such that

$$0 \leq U(t)u^o \leq U(t)z^o \tag{5.29}$$

where $\{U(t)z^o, t \in \mathbb{R}_+\}$ is monotone nonincreasing and converges to ϕ.

On the other hand, since $u^o \neq 0$, for any $\nu > 0$ we know by the strong positivity of $U(\nu)$ that $U(\nu)u^o \gg 0$, and it must satisfy the boundary conditions (5.7b).

We can select a lower solution $\underline{\phi}_\gamma(x)$ in the family (5.28) such that

$$\underline{\phi}_\gamma \leq U(\nu)u^o$$

As a consequence of the monotonicity of $U(t)$ in X_+,

$$U(t)\underline{\phi}_\gamma \leq U(t+\nu)u^o, \quad \text{for any} \quad t \in \mathbb{R}_+, \tag{5.30}$$

where $U(t)\underline{\phi}_\gamma$ is monotone nondecreasing, and converges to ϕ.

We can then state that

$$\lim_{t\to\infty} U(t)u^o = \phi \tag{5.31}$$

Due to the monotonicity of $\left\{U(t)\underline{\phi}_\gamma, t \in \mathbb{R}_+\right\}$ and of $\{U(t)z^o, t \in \mathbb{R}_+\}$ we obtain the GAS of $\phi \gg 0$ in $X_+ - \{0\}$.

We can remark here that in order to prove Theorem 5.5 we have heavily used the information already available for f in the ODE system (4.18).

Diffusion has only shifted the threshold parameter by shifting the loss terms in f by the first eigenvalue of the Laplace operator subject to the required boundary conditions.

This is strictly due to the quasimonotone (positive feedback) structure of our system.

We shall exploit further this "structural stability" of the ODE system with respect to diffusion by showing that this holds also in the periodic case analyzed in Section 4.5.

Diffusion also in this case will only drift the threshold condition (4.34) for existence of a nontrivial periodic endemic state.

(Extensions and refinements of the above results can be found in [33, 109]).

5.3. The periodic case

We shall consider now the case in which the force of infection seasonally depends on time, i.e. periodically [59].

The corresponding space homogeneous case has been analyzed in Section 4.5 (see system (4.28)).

We have seen already there that the analysis of the periodic solutions of a periodic system with the same period, reduces to a discrete time autonomous system. The same holds for the PDE system with suitable modifications.

If we introduce diffusion in system (4.28) we have, as we did for system (5.7),

$$\text{(5.32)} \qquad \frac{\partial}{\partial t}u(x;t) = D\Delta u(x;t) + f(t;u(x;t)) \ ,$$

$(u = (u_1, u_2)^T)$), for $x \in \Omega$, and $t > 0$, subject to the standard boundary conditions

$$\text{(5.32}b\text{)} \qquad B_i u_i(x;t) := \frac{\partial}{\partial \nu}u_i(x;t) + \alpha_i(x)u_i(x;t) = 0 \ , \quad i = 1,2 \ ,$$

for $x \in \partial\Omega$, and $t > 0$.

On Ω and α we shall have the same assumptions as in the autonomous case (5.7), (5.7b).

As far as f is concerned we shall take, as for system (4.28)

$$(t,z) = (t,z_1,z_2) \in \mathbb{R} \times \mathbb{K} \to f(t;z) = \begin{pmatrix} f_1(t;z) \\ \\ f_2(t;z) \end{pmatrix}$$

where

$$\begin{pmatrix} f_1(t;z) \\ \\ f_2(t;z) \end{pmatrix} := \begin{pmatrix} -a_{11}z_1 + a_{12}z_2 \\ \\ p(t)h(z_1) - a_{22}z_2 \end{pmatrix} \ ;$$

here p is a strictly positive Hölder continuous function, ω-periodic in $\mathbb{R}$, with $\omega \geq 0$:

$$p(t) = p(t+\omega), \quad t \in \mathbb{R} \ .$$

On h we shall make the same assumptions (i)-(v) as in Section 4.5.

Thus conditions (F1)-(F5) are again satisfied at any $t \in \mathbb{R}$ for f, which in addition will now be ω-periodic in time

(F6) $f(t;\cdot) = f(t+\omega;\cdot), \quad t \in \mathbb{R}$.

Again we shall suppose that all conditions are satisfied for the existence and uniqueness of a classical solution u of problem (5.32), (5.32b) subject to an initial condition

$$u(\cdot\,;s) = u(s) \in X_+, \quad s \in \mathbb{R} \tag{5.33}$$

(see [2]), which can be extended, because of (F3), to all $t \geq s$.

For any $(s,t) \in \Sigma := \{(s,t) \in \mathbb{R}^2 \mid s \leq t\}$, we shall denote by $U(t,s)u(s)$ the unique solution of problem (5.32), (5.32b) subject to the initial condition (5.33).

Due to the listed properties on f, we can confirm here for

$$\{U(t,s),\ (t,s) \in \Sigma\}$$

similar properties as for the autonomous case.

In particular we report here the following

(5.34) (Monotonicity)

$$\text{for any} \quad u^o, v^o \in X_+,: u^o \leq v^o \Longrightarrow U(t,s)u^o \leq U(t,s)v^o, \quad (s,t) \in \Sigma\ .$$

(5.35) (Positivity)

$$U(t,s)X_+ \subset X_+\ , \qquad (s,t) \in \Sigma$$

In the sequel we shall denote by U the monodromy operator (Poincaré map) of system (5.32), (5.32b).

$$U := U(\omega, 0)\ . \tag{5.36}$$

Under the assumptions (F1)-(F6) it can be shown that U is strongly positive and monotone (increasing).

It is well known that if $\{u(t), t \in \mathbb{R}\}$ is a solution of system (5.32), (5.32b) then the following two propositions are equivalent

(a) $\{u(t), t \in \mathbb{R}\}$ is a periodic solution with period ω of system (5.32), (5.32b);

(b) for any $s \in \mathbb{R}$, $u(s)$ is a fixed point of the monodromy operator U,

(usually in (b) we shall take $s = 0$) [3, 137].

As far as the stability of a periodic solution with period ω is concerned we may analogously state the equivalence of the following two propositions (see [59])

(a') $\{u(t)\,,\ t \in \mathbb{R}\}$ is a stable (asymptotically stable) periodic solution with period ω of system (5.32), (5.32b);

(b') $u(0)$ is a stable (asymptotically stable) fixed point with respect to the monodromy operator U.

By this we mean the following.

Let $\phi \in X_+$ be a fixed point of U. We say that ϕ is stable with respect to U if and only if for any choice of the number $\epsilon > 0$ we can find a number $\delta > 0$ (depending upon ϵ) such that whenever for $u^o \in X_+$, $\|u^o - \phi\| < \delta$, we have $\|U^n u^o - \phi\| < \epsilon$ for any integer $n \in \mathbb{N} - \{0\}$.

We say that ϕ is globally asymptotically stable in $D \subset X_+$ with respect to U if and only if ϕ is stable in D with respect to U, and, for any choice of $u^o \in D$, $\lim_{n\to\infty} \|U^n u^o - \phi\| = 0$.

As we did for the autonomous case, in order to analyze the asymptotic behavior of system (5.32), (5.32b) we shall take advantage of all information available for the corresponding ODE system (4.28). The key results are contained in Theorem 4.5.

Please note that whenever applicable we shall keep the same notations as in Section 5.2.

Theorem 5.6. *If*

$$\theta_1 := \frac{a_{12} p_{\max} h'(0)}{(a_{11} + d_1 \lambda_m)\,(a_{22} + d_2 \lambda_m)} < 1 \tag{5.37}$$

then the trivial solution is GAS in X_+ for system (5.32), (5.32b). If

$$\theta_2 := \frac{a_{12}\, p_{\min}\, h'(0)}{(a_{11} + d_1 \lambda_M)\,(a_{22} + d_2 \lambda_M)} > 1 \tag{5.38}$$

then the trivial solution is unstable for system (5.32), (5.32b).

Proof. To prove the first part of the theorem, we may proceed as in the proof of Theorem 5.1 by taking into account that (4.36) holds true.
For the second part we proceed as in Theorem 5.2, by taking into account that now (4.39) holds true.

5.3.1. Existence and stability of a nontrivial periodic endemic state

In this Section we shall prove that under condition (5.38), $\theta_2 > 1$, a unique nontrivial periodic endemic state exists for system (5.32), (5.32b).

In order to do so we shall apply an extension to the periodic case of the method of lower solutions and upper solutions as discussed in [2, 133, 136].

Definition 5.7. We shall say that $\underline{u}(x;t)$ is an ω-lower solution of system (5.32), (5.32b) in $(0,+\infty)$ iff $\underline{u}(x;t)$ is a classical solution

$$\underline{u} \in C^{2,1}\left(\Omega \times (0,+\infty), \mathbb{R}^2\right) \cap C^{1,0}\left(\overline{\Omega} \times (0,+\infty), \mathbb{R}^2\right)$$

of the following problem.

$$(5.39) \qquad \frac{\partial}{\partial t}\underline{u}(x;t) \leq D\Delta\underline{u}(x;t) + f(t,\underline{u}(x;t))$$

in $\Omega \times (0,+\infty)$, with boundary conditions

$$(5.39b) \qquad \frac{\partial}{\partial \nu}\underline{u}_i(x;t) + \alpha_i(x;t)\underline{u}_i(x;t) \leq 0 \ , \qquad i = 1,2$$

in $\partial\Omega \times (0,+\infty)$; and

$$(5.39\omega) \qquad \underline{u}(x;0) \leq \underline{u}(x;\omega) \ , \qquad \text{in } \overline{\Omega}$$

We speak of a strict ω-lower solution if any of the inequalities in (5.39), (5.39b), (5.39ω) is strict.

The notion of ω-upper solution (or strict ω-upper solution) is given by reversing the above inequalities.

The following lemmas hold [59].

Lemma 5.8. *If $\theta_2 > 1$ (see (5.38)) the following system*

$$\frac{d}{dt}\underline{w}(t) = -\lambda_M D\underline{w}(t) + f(t;\underline{w}(t)) \ , \qquad t \in \mathbb{R} \tag{5.40}$$

admits a nontrivial periodic solution $\underline{w}(t)$, $t \in \mathbb{R}$, which is GAS in $\mathbb{K} - \{0\}$. Moreover, for any $\epsilon \in (0,1)$

$$\underline{u}_\epsilon(x;t) := \epsilon\phi_M(x)\,\underline{w}(t) \ , \qquad \text{in} \quad \overline{\Omega} \times \mathbb{R}$$

is a strict ω-lower solution for system (5.32), (5.32b).

Proof. The first part of the lemma follows from the second part of Theorem 5.6. For the second part, as a consequence of (F5), we have that

$$\begin{aligned}\frac{\partial}{\partial t}\underline{u}(x;t) &= \epsilon\phi_M(x)\frac{dw}{dt} = -\lambda_M D\left(\epsilon\phi_M(x)w(t)\right) + \epsilon\phi_M(x)f(t,\underline{w}(t)) \\ &\ll D\Delta\underline{u}(x;t) + f(t,\underline{u}(x;t))\end{aligned}$$

in $\Omega \times \mathbb{R}$.

Lemma 5.9. *If $\theta_2 > 1$, the ODE system (4.28) admits a nontrivial ω-periodic solution $z(t), t \in \mathbb{R}$, which is GAS in $\mathbb{K} - \{0\}$. It is an ω-upper solution of system (5.32), (5.32b).*

Proof. It is a straightforward consequence of Theorem 4.5.

By extending to this case the techniques used in Appendix B it can be shown that the following theorem holds [59]

Theorem 5.10.

(a) *If $\underline{u}(x;t)$ is an ω-lower solution of system (5.32), (5.32b) in $(0,+\infty)$, the sequence*

$$\{U^n\underline{u}(\cdot\,;0); \quad n \in \mathbb{N}\}$$

is monotone nondecreasing in X_+.

(b) *On the other hand, if $\overline{u}(x;t)$ is an ω-upper solution of system (5.32), (5.32b) in $(0,+\infty)$, the sequence*

$$\{U^n\overline{u}(\cdot\,;0); \quad n \in \mathbb{N}\}$$

is monotone nonincreasing in X_+.

(c) *A minimal fixed point $\phi_- \in X_+$, and a maximal fixed point $\phi_+ \in X_+$, exist for the monodromy operator U*

$$U\phi_- = \phi_- \quad \text{and} \quad U\phi_+ = \phi_+$$

$$\text{such that} \quad 0 \ll \phi_- \le \phi_+ \ .$$

Moreover if $\underline{u}$ and $\overline{u}$ are as in (a) and (b), respectively,

$$\lim_n d\left(U^n \underline{u}(\cdot\,;0), [\phi_-, +\infty)\right) = 0$$

$$\lim_n d\left(U^n \overline{u}(\cdot\,;0), [0, \phi_+]\right) = 0$$

This implies the following main theorem.

Theorem 5.11. *If $\theta_2 > 1$ (see (5.38)), then system (5.32), (5.32b) admits a unique nontrivial periodic solution (with positive components) which is GAS in $X_+ - \{0\}$.*

Proof. Due to Theorem 5.10, a minimal fixed point ϕ_- and a maximal fixed point ϕ_+ exist in X_+ for the monodromy operator U of system (5.32), (5.32b).

Under conditions (F4) and (F5) on f, it can be also shown [59] that U is strongly concave. Thus [139; p.188] U cannot have more than one nontrivial fixed point in X_+.

Hence $0 \ll \phi_- = \phi_+$.

Let us denote by ϕ such a fixed point.

Now for any choice of $u^o \in X_+ - \{0\}$ due to condition (F3) on f we can find a $\xi \in \overset{\circ}{\mathbb{K}}$ such that $0 \le u^o \le \xi$, and $f(t;\xi) \le 0$; thus ξ is an ω-upper solution of system (5.32), (5.32b). By the monotonicity of U we have

$$0 \le U^n u^o \le U^n \xi$$

and by (c) in Theorem 5.10,

$$\lim_n d\left(U^n u^o, [0, \phi]\right) = 0$$

On the other hand, since $u^o \neq 0$, we know by the strong positivity of U that $Uu^o \gg 0$, and it must satisfy the boundary conditions (5.32b). By Lemma 5.8, we can then find an $\epsilon > 0$ such that

$$\underline{u}^\epsilon \le Uu^o \ .$$

As a consequence of the monotonicity of U in X_+ we shall have

$$U^n \underline{u}^\epsilon \leq U^{n+1} u^o$$

and by (c) in Theorem 5.10,

$$\lim_n d\,(U^n u^o, [\phi, +\infty)) = 0$$

Altogether then we have proved the global attractivity of ϕ in X_+

$$\lim_n U^n u^o = \phi$$

As far as the stability of ϕ is concerned, we may observe that in the proof for any choice of $u^o \in X_+ - \{0\}$ we have found an ω-lower solution $\underline{u}$ and a ω-upper solution $\overline{u}$ of system (5.32), (5.32b) such that

$$\underline{u}(\cdot, 0) \leq u^o \leq \overline{u}(\cdot, 0)\ .$$

By Theorem 5.10, $U^n \underline{u}(\cdot, 0)$ converges monotonically nondecreasing to ϕ, while $U^n \overline{u}(\cdot, 0)$ converges monotonically nonincreasing to ϕ. This implies with the global attractivity, the GAS of ϕ in $X_+ - \{0\}$.

5.4. Saddle point behavior [58]

We consider now the reaction-diffusion system (5.7), (5.7b) of Section 5.2 with f defined by (5.8) and g satisfying conditions (i)-(iv) and (v') of Section 4.4, so that Theorem 4.3 applies.

Since f still satisfies conditions (F1)-(F3) of Section 5.2, we may still claim properties (5.10) and (5.11) for the family of evolution operators $\{U(t),\ t \in \mathbb{R}_+\}$.

We distinguish two main cases with respect to the boundary conditions (5.7b); homogeneous Neumann boundary conditions ($\alpha_1 = \alpha_2 = 0$) and general third type boundary conditions (which include the case of homogeneous Dirichlet boundary conditions).

5.4.1. Homogeneous Neumann boundary conditions

It is clear that Theorem 4.3 still holds true for the PDE system as far as cases a) and b) are concerned.

On the other hand, under the assumptions of Theorem 4.3 c) it is still true that the only spatially homogeneous equilibrium solutions of the PDE system (5.7) are given by those of the corresponding ODE system (4.18) that we have denoted by $0, z^*, z^{**}$.

Following [68, 108] it can be shown that two sets M'_+ and M'_- exist, both of them contained in X_+, respectively called stable and unstable manifold of z^* such that if $u^o \in M'_+$, then $U(t)u^o \longrightarrow z^*$ (as $t \longrightarrow +\infty$), while if $u^o \in M'_-$, then a unique backward extension $\overline{u}(t; u^o)$, $t \in \mathbb{R}_- := (-\infty, 0]$ exists for the solution $\{U(t)u^o\,,\ t \in \mathbb{R}_+\}$, such that $\overline{u}(t; u^o) \longrightarrow z^*$ (as $t \longrightarrow -\infty$). The structure of these manifolds it is not clear in general but, in this case, we can establish a structural correspondence between M_+, M_- in Theorem 4.3 c) and M'_+, M'_-. In fact the following theorem has been established [58]:

Theorem 5.12. *Let the assumptions of Theorem 4.3 c) be satisfied. Then $M_+ \subset M'_+$; moreover, if $u^o \in X_+$ is such that two elements ξ^o, $\eta^o \in M_+$ exist for which*

$$\xi^o \le u^o(x) \le \eta^o \,, \qquad x \in \overline{\Omega} \tag{5.41}$$

then

$$\lim_{t \to +\infty} \| U(t)u^o - z^* \|_X = 0 \,.$$

Also $M_- \subset M'_-$; moreover, if β is the negative eigenvalue of $J_f(z^*)$, then two numbers ρ, $\gamma > 0$ exist such that for any $u^o \in B_\rho(z^*)$, if $t_0 > 0$ is such that for any $t \in [0, t_0]$, $U(t)u^o \in B_\rho(z^*)$, then

$$dist(U(t)u^o\,,\ M'_-) < \gamma \exp(\beta t) \,, \qquad t \in [0, t_0] \,. \tag{5.42}$$

Proof. The first part of the theorem follows from the following remark.

If initially $u^o \in X_+$ satisfies condition (5.41) then

$$V(t)\,\xi^o \le U(t)u^o \le V(t)\,\eta^0 \,, \qquad t \in \mathbb{R}_+ \,, \tag{5.43}$$

where $\{V(t)\,,\ t \in \mathbb{R}_+\}$ is the family of evolution operators of the ODE system (4.18) and $\{U(t)\,,\ t \in \mathbb{R}_+\}$ is the family of evolution operators of the PDE system (5.7), (5.7b).

The second part follows from Theorem 3.3 in [68] and from the properties of the spectrum $\sigma(A + J_f(z^*))$ [103].

We may remark here that if we denote by

$$\mathcal{R}'_2 := \{u \in X_+ \mid z^* \le u\}$$

and

$$\mathcal{R}'_4 := \{u \in X_+ \mid u \le z^*\}$$

then, thanks to Theorem 5.12 we can state that

$$M'_+ \cap (\mathcal{R}'_2 \cup \mathcal{R}'_4) = \{z^*\} .$$

Further properties of the domains of attraction $dom(0)$ and $dom(z^{**})$ for the parabolic system (5.7) subject to homogeneous Neumann boundary conditions can be obtained as a consequence of Theorem 5.12 and in particular of (5.43).

Proposition 5.13. *Under the assumptions of Theorem 5.12, if $u^o \in X_+$ is majorized by a $\eta^o \in dom(0)$, then*

$$\lim_{t \to +\infty} \|U(t)u^o\|_X = 0 .$$

If $u^o \in X_+$ is minorized by a $\xi^o \in dom(z^{**})$, then

$$\lim_{t \to +\infty} \|U(t)u^o - z^{**}\|_X = 0 .$$

Up to now we have noticed that with homogeneous Neumann boundary conditions the equilibria of the ODE system are preserved in the corresponding PDE system; it is also true that the asymptotic behavior of solutions which arise from initial conditions which are comparable with elements of $dom(0)$ and $dom(z^{**})$ is reproduced in the PDE system. But if some $u^o \in X_+$ cannot be compared with elements of either $dom(0)$ or $dom(z^{**})$, then it is rather difficult to predict analytically the qualitative behavior of the solution $U(t)u^o$.

Numerical simulations have been carried out to explore the behavior of solutions of the PDE system, which do not fall under the assumptions of Proposition 5.13 . In both Figs. 5.1 and 5.2 homogeneous Neumann boundary conditions are assumed, with different diffusion coefficients. In both cases the initial condition $u^o \in X_+$ is such that its range of values in $\mathbb{K}$ "crosses" the

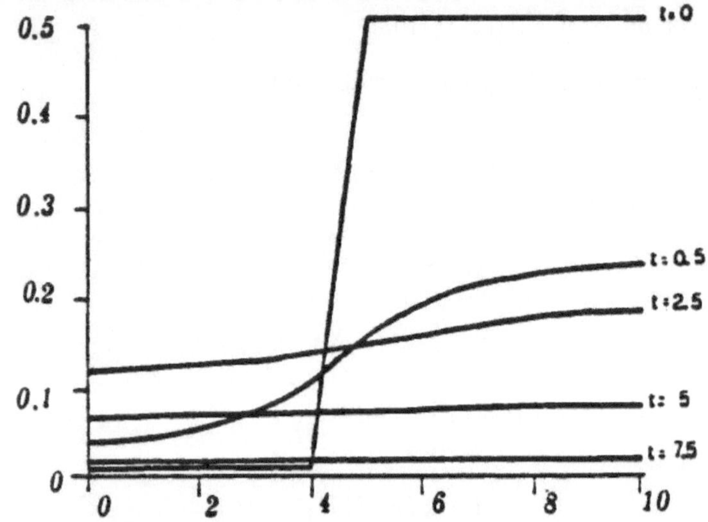

Fig. 5.1. Homogeneous Neumann boundary conditions; $d_1 = d_2 = 0.1$; $a_{11}a_{22} = 1$; $a_{12} = 4$; $\Delta t = 0.5$; $\Theta = 0$ [58].

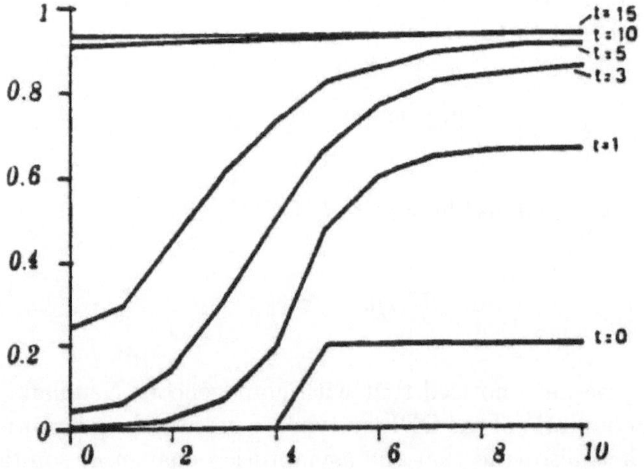

Fig. 5.2. Homogeneous Neumann boundary conditions; $d_1 = d_2 = 0.01$; $a_{11}a_{22} = 1$; $a_{12} = 4$; $\Delta t = 0.5$; $\Theta = 0$ [58].

manifold M_+ of the ODE system; it belongs in part to $dom(0)$ and in part to $dom(z^{**})$. Fig. 5.1 shows an eventual tendency of the solution toward 0, while Fig. 5.2 shows an eventual tendency toward z^{**} [58].

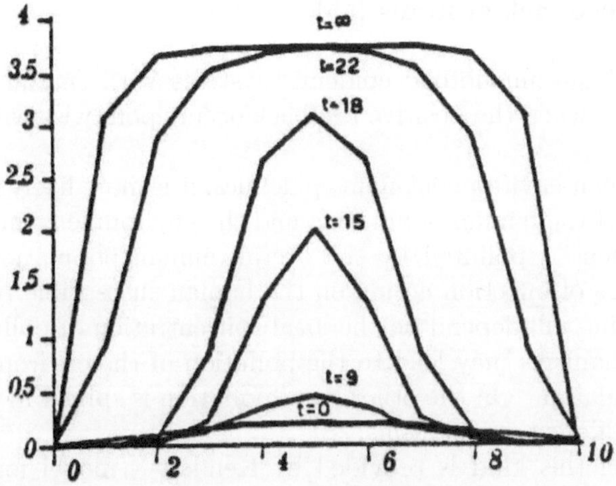

Fig. 5.3. Homogeneous Dirichlet boundary conditions;
$d_1 = d_2 = 0.01$; $a_{11}a_{22} = 1$; $a_{12} = 4$; $\Delta t = 0.5$; $\Theta = 0$ [58].

5.4.2. Boundary conditions of the third type

When the coefficients α_i in (5.7b) are not both identically zero, the only spatially homogeneous equilibrium solution of system (5.7), (5.7b) is the trivial one.

The analysis in this case is rather complicated technically, and goes beyond the scope of this presentation. The interested reader may refer to [58].

Anyway, Fig. 5.3 illustrates the asymptotic behavior of a solution of our PDE system subject to homogeneous Dirichlet boundary conditions, and an initial condition which again "crosses" in $\mathbb{K}$ the manifold M_+. It shows the eventual tendency of the solution toward a nontrivial, spatially nonhomogeneous equilibrium state; even though, for short times, the trivial solution "attracts" the "tails" of the initial distribution $u^o(x)$.

5.5. Boundary feedback systems [55]

In Section 5.2 quasimonotone epidemic systems with spatial structure have been analyzed, when the positive feedback occurs pointwise with respect to space.

Actually, for man-environment-man epidemics, it is more likely to assume global interaction of the human population and the environment, in the sense that the environment is polluted by the overall human population; at any point then the force of infection acting on the human susceptible population present at that point will depend on the local concentration of pollution.

Different mechanisms may lead to the pollution of the environment due to the human population. On the other hand pollution is spread in the whole habitat again by different mechanisms.

An example of this kind is provided by Kendall 's model for the SIR model with spatial structure [131]. Multigroup models such as those discussed in Sections 2.3.4 and 4.6 may also be considered as discretized versions of Kendall like models [22, 29, 142].

In this section we shall present the case in which the reservoir of the pollutant generated by the human population is spatially separated from the habitat by a boundary through which the positive feedback occurs.

A model of this kind has been proposed as an extension of the model already presented in Sect. 4.3.4 for fecal-oral transmitted infections in Mediterranean coastal regions.

For this kind of epidemics the infectious agent is multiplied by the infective human population and then sent to the sea through the sewage system; because of the peculiar eating habits of the population of these regions, the agent may return via some diffusion-transport mechanism to any point of the habitat, where the infection process is restarted.

The mathematical model is based on the following system of evolution equations [55]:

$$
\left\{
\begin{aligned}
\frac{\partial}{\partial t}u_1(x;t) &= \Delta u_1(x;t) - a_{11}u_1(x;t) \\
\frac{\partial}{\partial t}u_2(x;t) &= -a_{22}u_2(x;t) + g(u_1(x;t))
\end{aligned}
\right.
\tag{5.44}
$$

in $\Omega \times (0, +\infty)$, subject to the following boundary condition

$$
\frac{\partial}{\partial \nu}u_1(x;t) + \alpha u_1(x;t) = \int_{\Omega} k(x,x')u_2(x';t)\,dx'
\tag{5.45}
$$

on $\partial\Omega \times (0, +\infty)$, and also subject to suitable initial conditions.

Here Δ is the usual Laplace operator modelling the random dispersal of the infectious agent in the habitat; the human infective population is supposed not to diffuse.

The meaning of the other terms is the same has for model (4.16) in Sect. 4.3.4.

In the boundary condition (5.45) the left hand side is the general boundary operator $B := \frac{\partial}{\partial \nu} + \alpha(\cdot)$ associated with the Laplace operator; on the right hand side the integral operator

$$H\left[u_2(\cdot,t)\right](x) := \int_\Omega k(x,x')u_2(x';t)\,dx' \tag{5.46}$$

describes boundary feedback mechanisms, according to which the infectious agent produced by the human infective population at time $t > 0$, at any point $x' \in \Omega$, is available, via the transfer kernel $k(x,x')$, at a point $x \in \partial\Omega$.

In model (5.46) delays are neglected and the feedback process is considered to be linear; extensions which overcome these simplifications would be welcome.

Clearly the boundary $\partial\Omega$ of the habitat Ω has to be divided into two disjoint parts : the sea shore Γ_1 through which the feedback mechanism may occur, and Γ_2 the boundary on the land, at which we may assume complete isolation.

In (5.45) the parameter $\alpha(x)$ denotes the rate at which the infectious agent is wasted away from the habitat into the sea at the sea shore.

Thus we shall assume that

$$\alpha(x), \quad k(x,\cdot) = 0, \qquad \text{for } \ x \in \Gamma_2\ .$$

We further assume that the habitat Ω is "epidemiologically" connected to its boundary by requesting that

$$\text{for any } \ x' \in \Omega \ \text{ there exists some } \ x \in \Gamma_1 \ \text{ such that } \ k(x,x') > 0 \tag{5.47}$$

This means that from any point of the habitat infective individuals contribute to polluting at least some point on the boundary (the sea shore).

We shall assume as we did in Section 5.2 that a_{11} and a_{22} are positive constants, and that the infection function g satisfies conditions (i)-(iv) in Section 4.4.

System (5.44), (5.45) has to be supplemented by an initial condition

(5.48) $$u(x;0) = u^o(x), \qquad x \in \Omega.$$

We shall assume, as usual, that all conditions are satisfied for the existence and uniqueness of a unique classical solution $\{u(t), t \geq 0\}$ of problem (5.44), (5.45), (5.48) [55]. By a classical solution in this case we mean

(5.49) $$u \in \left(C^{2,1}(\Omega \times (0,+\infty), \mathbb{R}) \cap C^{1,0}\left(\overline{\Omega} \times (0,+\infty), \mathbb{R}\right)\right)$$
$$\times C^{0,1}\left(\overline{\Omega} \times (0,+\infty), \mathbb{R}\right).$$

We shall then proceed as in Sect. 5.2 by exploring the existence of non-trivial time-independent endemic states for system (5.44), (5.45).

Let us observe first that an equilibrium solution of system (5.44), (5.45) is a solution of the following nonlinear elliptic system

(5.50) $$\begin{cases} \Delta u_1 - a_{11}u_1 = 0 \ , & \text{in } \ \Omega \\ Bu_1 = Hu_2 \ , & \text{on } \ \partial\Omega \\ -a_{22}u_2 + g(u_1) = 0 \ , & \text{in } \ \Omega, \end{cases}$$

which can obviously be reduced to

(5.51) $$\begin{cases} \Delta u_1 - a_{11}u_1 = 0 \ , & \text{in } \ \Omega \\ Bu_1 = \dfrac{1}{a_{22}}H[g(u_1)] \ , & \text{on } \ \partial\Omega. \end{cases}$$

If we denote by $S_\omega(k,h)$ the unique (classical) solution of the following linear elliptic problem ($\omega \geq 0$ is a real parameter)

(5.52) $$\begin{cases} -\Delta v + (a_{11} + \omega)v = k \ , & \text{in } \ \Omega \\ Bv = h \ , & \text{on } \ \partial\Omega \end{cases}$$

given $k \in C(\overline{\Omega}), h \in C(\partial\Omega)$, (see [1]), we can formally solve system (5.51) to get

(5.53) $$u_1 = S_\omega \left(\omega u_1, \frac{1}{a_{22}} H[g(u_1)] \right)$$

which shows that our problem is equivalent to a fixed point problem for the nonlinear operator

(5.54) $$\Phi_\omega(w) := S_\omega \left(\omega w, \frac{1}{a_{22}} H[g(w)] \right).$$

The solution is meant in the classical sense $(w \in C^2(\Omega) \cap C^1(\partial\Omega))$.

Under the assumptions on g, a_{22}, and H, the nonlinear operator Φ_ω is strictly monotone increasing , strictly concave and compact, for any $\omega \geq 0$.

In [62] it is shown that the solution of problem (5.51) is intimately related to the solution of the linear eigenvalue problem

(5.55) $$\begin{cases} \Delta v - a_{11} v = 0 \ , & \text{in} \quad \Omega \\ Bv = \xi H v \ , & \text{on} \quad \partial\Omega \end{cases}$$

(here Δv has to be interpreted in the distributional sense), with $\xi > 0$ and $v \in C(\overline{\Omega}), v \geq 0$.

If we make use of the solution operator S_ω of problem (5.52), problem (5.55) is equivalent to

(5.56) $$v = S_\xi(\xi v, \xi H v) = \xi S_\xi(v, Hv) =: Z_\xi(v)$$

which is a fixed point problem for the linear operator Z_ξ in $C(\overline{\Omega})$.

The following proposition is a standard result from the theory of linear strictly positive compact operators in ordered Banach spaces (see e.g. [197]; see also Theorem B.9 in Appendix B).

Proposition 5.14.

a) *For $\xi \in \mathbb{R}_+ - \{0\}, Z_\xi$ is a strictly positive compact linear operator.*

b) *For $\xi \in \mathbb{R}_+ - \{0\}, \quad spr\ Z_\xi > 0$*

c) *For $\xi \in \mathbb{R}_+ - \{0\}$ a unique $v \in C(\overline{\Omega}), v \gg 0, \|v\| = 1$ exists such that*

$$Z_\xi v = (spr\ Z_\xi) v$$

d) *spr* Z_ξ *is the only eigenvalue of* Z_ξ *with a positive eigenvector.*

Note that here *spr* Z_ξ denotes the spectral radius of Z_ξ.

As (5.55) is equivalent to (5.56) it is sufficient by Proposition 5.14 to find a $\xi > 0$ such that *spr* $Z_\xi = 1$, and to show that ξ is unique.

The first part follows from the following lemma.

Lemma 5.15.

a) *spr* $Z_\xi \to 0$ *as* $\xi \to 0$

b) *spr* $Z_\xi \to +\infty$ *as* $\xi \to +\infty$

c) *spr* Z_ξ *depends continuously on* $\xi > 0$.

Lemma 5.15 together with Proposition 5.14 imply now that a solution ξ, v exists to (5.55) with $\xi \in \mathbb{R}_+ - \{0\}, v \in C(\overline{\Omega}), v \gg 0$.

Uniqueness follows from the following lemma and Prop. 5.14 c), d).

Lemma 5.16. *Let* $v, \tilde{v} \in C(\overline{\Omega}), v, \tilde{v} > 0$ *such that* $\Delta v, \Delta\tilde{v} \in C(\overline{\Omega})$ *and*

$$\Delta v \leq a_{11} v \ , \qquad in \quad \Omega$$

$$Bv \geq \xi H v \ , \qquad on \quad \partial\Omega$$

and

$$\Delta\tilde{v} \geq a_{11}\tilde{v} \ , \qquad in \quad \Omega$$

$$B\tilde{v} \leq \eta H \tilde{v} \ , \qquad on \quad \partial\Omega$$

with $\xi, \eta \in \mathbb{R}_+$. *Then* $\eta \leq \xi$.

We have thus proven the following

Theorem 5.17. *There exists a unique pair* $\xi \in \mathbb{R}_+ - \{0\}, v \in C(\overline{\Omega}), v \geq 0, \|v\| = 1$ *which solves problem (5.55). Moreover* $v \gg 0$.

We are ready to state the main theorem about the existence of a nontrivial endemic state. For the proof we refer to [63].

Theorem 5.18. *Let* ξ, v *be the unique pair from the Theorem 5.17 solving (5.55).*

a) *If* $g'(0)/a_{22} \leq \xi$, *then* $u_1 = 0, u_2 = 0$ *is the only nonnegative equilibrium solution for (5.44), (5.45).*

b) *If $g'(0)/a_{22} > \xi$, then there exists a unique nonnegative nontrivial equilibrium solution $u_1, u_2 \in C(\overline{\Omega})$ for (5.44), (5.45). Moreover $u_1(x), u_2(x) > 0$ for any $x \in \overline{\Omega}$.*

Theorem 5.18 suggests that, if $g'(0)/a_{22} \leq \xi$ with ξ from Theorem 5.17, then the epidemic described by system (5.44), (5.45) dies out, whereas, if $g'(0)\ a_{22} > \xi$ it tends to a nontrivial endemic state.

This statement has still to be proven rigorously.

Thus Theorem 5.18 can be seen as a "threshold theorem" for the epidemic system. It is then interesting to estimate ξ from the data of model (5.44), (5.45).

A general answer to this problem has not been given yet. In the particular case in which the washout rate $\alpha(x)$ is identically zero, an explicit threshold condition can be found for a general geometric structure of Ω. In this case the following theorem holds [63].

Theorem 5.19. *Let $\alpha \equiv 0$ in (5.45).*

a) *If*

$$\frac{g'(0)}{a_{11}a_{22}} \sup_{x' \in \Omega} \int_{\partial\Omega} k(x,x')\, d\sigma(x) \leq 1$$

then no nontrivial endemic state exists.

b) *If*

$$\frac{g'(0)}{a_{11}a_{22}} \inf_{x' \in \Omega} \int_{\partial\Omega} k(x,x')\, d\sigma(x) > 1$$

then there exists a unique nontrivial endemic state.

Theorem 5.19 relates what we may call the "threshold parameter"

$$\tilde{\theta} := \frac{g'(0)}{a_{11}a_{22}} \sup_{x' \in \Omega} \int_{\partial\Omega} k(x,x')\, d\sigma(x) \tag{5.57}$$

to the biological parameters of the system.

The problem of identifying these parameters arises; the identification of the kernel $k(x, x')$ has been faced in [141].

It is not difficult to remark that $\tilde{\theta}$ in (5.57) is an intuitive generalization of the parameter θ in (4.20); the multiplication rate a_{12} of the infectious agent due to the human infective population is naturally changed into the overall production rate $\sup_{x' \in \Omega} \int_{\partial\Omega} k(x,x')\, d\sigma(x)$.

This can be seen as a further test about the "robustness" of the basic model (4.18) with respect to the introduction of a spatial structure in the model itself, as already shown in Sect. 5.2.

This adds to the robustness with respect to time periodicity shown in Sect. 4.5 and 5.3 .

We wish to emphasize that this is due to the quasimonotonicity of system (4.18) together with the strict concavity of the force of infection g.

Numerical simulations of system (5.44), (5.45) subject to different initial conditions have been carried out in [13]. Figs. 5.4 and 5.5 show the behavior of the integrals

$$U_i(t) = \int_\Omega u_i(x;t)\,dx\ , \qquad i = 1,2$$

as functions of time. The force of infection g has been taken of the form

$$g(u_1) = \beta\,\frac{u_1}{\gamma u_1 + 1}\ .$$

In Fig. 5.4, $\beta = 0.02$ for which a behavior "below threshold" is obtained; Fig 5.5, with $\beta = 50$ shows a behavior "above threshold".

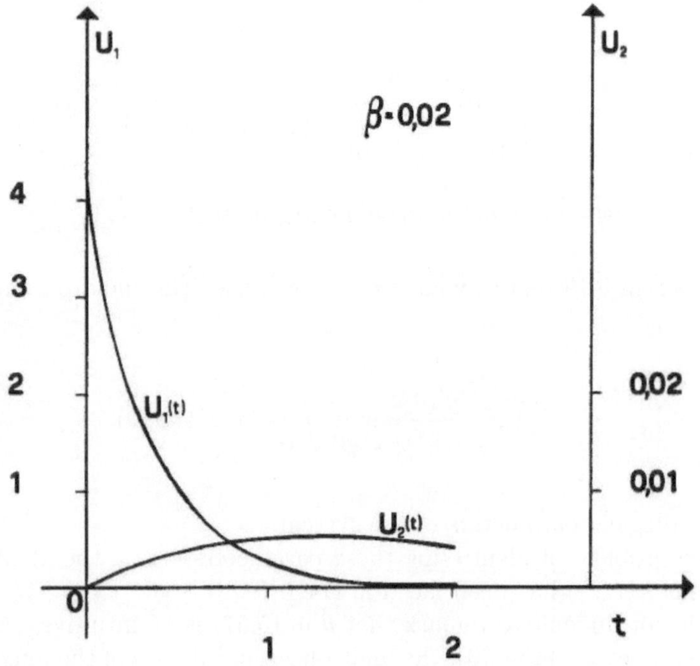

Fig. 5.4. $\beta = 0.02$: Asymptotic extinction of the epidemic [13].

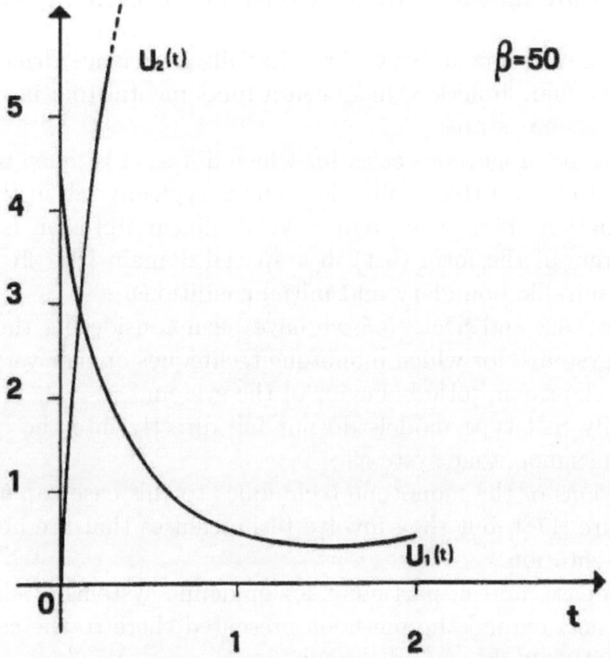

Fig. 5.5. $\beta = 50$: Asymptotic tendency to a nontrivial endemic state [13].

5.6. Lyapunov methods for spatially heterogeneous systems

We have seen how in Sect. 2 the LaSalle Invariance Principle based on a suitable Lyapunov function, has been a fundamental tool in proving GAS of nontrivial endemic states.

In space heterogeneous case, for which diffusion is taken into account, we have seen (Sect. 5.1) that epidemic systems typically fall in the more general class of reaction-diffusion systems. When linear diffusion is considered we have a system of the form (5.1) in a spatial domain $\Omega \subset \mathbb{R}^m$, $m = 1, 2, 3$, subject to suitable boundary and initial conditions.

In Sect. 5.2 and Sect. 5.5 we have been considering the case of quasi-monotone systems, for which monotone techniques appear very fruitful in the analysis of the asymptotic behavior of the system.

Actually SIR type models do not fall directly into the class of positive feedback quasimonotone systems.

Extensions of the monotone techniques to this case can still be found in the literature [178], but they involve technicalities that are beyond the scope of this presentation.

In this case, and in particular for epidemic systems of class A and B of Sect.2, we may extend the methods presented there to the case with spatial diffusion, with suitable modifications.

An earlier use of Lyapunov functions to study reaction-diffusion systems can be found in [146, 185, 192, 221] and others, and a unified presentation of the results is given in [107].

The most interesting results had been obtained for spatially independent steady states, which are typically implied by zero Neumann boundary conditions, while a unified treatment for more general boundary conditions and for the case of spatially dependent steady states based on Lyapunov functions is still missing.

Here we shall refer in fact to the case of epidemic systems with diffusion which fall in the general structure discussed in Sect.2, subject to zero Neumann boundary conditions. The results, which were presented in [48], are based on the approach due to Redheffer and Walter [187] who have analyzed by Lyapunov methods reaction-diffusion systems of the Volterra predator-prey type, of which our general structure is a possible extension (see also [64, 147]).

We shall consider the following semilinear parabolic system

$$\frac{\partial u}{\partial t}(x;t) = D\Delta\, u(x;t) + f(u(x;t)) \ , \tag{5.58}$$

$u := (u_1, \cdots, u_n)^T$, in $\Omega \times (0, +\infty)$, subject to zero Neumann boundary conditions

$$\frac{\partial u}{\partial \nu}(x;t) = 0 \ , \qquad \text{on } \partial\Omega \times (0, +\infty) \tag{5.58b}$$

and initial conditions

$$(5.58o) \qquad u_i(x;0) = u_i^o(x)\ , \qquad i = 1, \ldots, n\ , \quad \text{in } \overline{\Omega}\,.$$

Here $u = (u_1, \ldots, u_n)^T$, and $f(z)$, $z \in \mathbb{R}_+^n$, is the interaction law among the species ($n \in \mathbb{N} - \{0\}$).

The present analysis refers to the case in which we have only "local" interactions between the species so that $f(z)$ will be the same as in the spatially homogeneous case (see e.g. Eqn. (5.2)).

The domain $\Omega \subset \mathbb{R}^m$ will be assumed to be an open, bounded, connected subset of the geographical space $\mathbb{R}^m$ ($m \in \mathbb{N} - \{0\}$), with a sufficiently smooth boundary $\partial\Omega$; $\partial/\partial\nu$ denotes the outward normal derivative; D is the diagonal matrix of the diffusion coefficients, and we shall assume that all $d_i > 0$, $i = 1, \ldots, n$ are equal so that $D = I_{n\times n}$. We are limiting our analysis to only the Laplace operator Δ for simplicity; it can be easily replaced to a general linear elliptic operator [187].

A relevant extension to different diffusion coefficients can be found in [49, 50].

We shall assume that all conditions are satisfied for existence and uniqueness of a classical solution u of problem (5.58), (5.58b), (5.58o) (see Appendix B, Section B.2), in a suitable time interval $[0, \overline{t})$.

By classical solution we mean that

$$(5.59) \qquad u \in C^{2,1}\left(\Omega \times (0, +\infty), \mathbb{R}^n\right) \cap C^{1,0}\left(\overline{\Omega} \times (0, +\infty), \mathbb{R}^n\right).$$

This suggests that we carry our analysis in the Banach space $X := C^2(\overline{\Omega})$ of real valued functions with continuous second order derivatives; its norm will be denoted by $\|\cdot\|_2$, defined in the usual way [108].

Since it can be shown (see later) that the solution $u(x;t)$ is bounded together with its derivatives up to the second order, it can be extended in t to the whole $\mathbb{R}_+$.

It makes sense then to consider the family of evolution operators $\{U(t),\ t \in \mathbb{R}_+\}$ such that whenever $u^o \in X$ (sufficiently smooth), $u(t) := U(t)\,u^o \in X$ is the solution of problem (5.58), (5.58b) with initial condition u^o. The trajectory of system (5.58), (5.58b) corresponding to the initial condition u^o is the subset $\Gamma_+(u^o) := \{U(t)\,u^o\,,\ t \in \mathbb{R}_+\}$.

It follows from well known results [108, 162] that if $\Gamma_+(u^o)$ is bounded in X, then it is precompact (its closure is compact; see Appendix B).

Given $u^o \in X$, the positive limit set $\omega(u^o)$ is defined as follows

$$\omega(u^o) := \Big\{ u \in X \;\Big|\; \text{there exists } (t_k)_{k\in\mathbb{N}}\,,\ t_k \geq 0\,, \qquad (5.60)$$

$$\text{s.t. } \lim_{k\to\infty} t_k = +\infty \text{ and } \lim_{k\to\infty} U(t_k)\,u^o = u \Big\}\,.$$

The following theorem is taken from the literature (see e.g. [108, 162] and Theorem B.34 in Appendix B):

Theorem 5.20. *If $\Gamma_+(u^o)$ is precompact in the Banach space X, then the positive limit set $\omega(u^o)$ is nonempty, compact, connected, and invariant; moreover*

$$\lim_{t\to+\infty} dist(U(t)\,u^o\,,\,\omega(u^o)) = 0 \qquad (5.61)$$

where the distance is measured in the norm $\|\cdot\|_2$ of X.

Let us consider our particular case in which

$$f : z \in \mathbb{R}^n_+ \longrightarrow f(z) = diag(z)\,(e + Az) + b(z) \qquad (5.62)$$

as defined in Sect. 2,

$e \in \mathbb{R}^n$	is a constant vector
$A = (a_{ij})_{i,j=1,\ldots,n}$	is a real constant matrix
$b(z) = c + Bz$	with
$c \in \mathbb{R}^n$	a constant vector
$B = (b_{ij})_{i,j=1,\ldots,n}$	a real constant matrix such that

$$b_{ij} \geq 0\,, \qquad i,j = 1,\ldots,n$$
$$b_{ii} = 0\,, \qquad i = 1,\ldots,n$$

We shall assume, as in Sect. 2, that a strictly positive equilibrium $z^* \gg 0$ exists for the ODE system

$$\frac{d}{dt} z = f(z) , \qquad t > 0 \tag{5.63}$$

i.e.

$$f(z^*) = 0 . \tag{5.64}$$

We may associate to z^* the classical Volterra-Goh Lyapunov function [96] as we did in Sect. 2

$$V(z) := \sum_{i=1}^{n} w_i \left(z_i - z_i^* - z_i^* \ln \frac{z_i}{z_i^*} \right) , \qquad z \in \mathbb{R}_+^{n*}$$

where $w_i > 0$, $i = 1, \ldots, n$ are real constants, and we have denoted by $\mathbb{R}_+^{n*} := \{z \in \mathbb{R}^n \mid z_i > 0, \ i = 1, \ldots, n\}$.

Clearly $V : \mathbb{R}_+^{n*} \longrightarrow \mathbb{R}_+$ is such that

i) $V(z^*) = 0$

ii) $V(z) > 0 , \qquad z \neq z^*$

iii) $V(z) \rightarrow +\infty$ for $z \rightarrow \partial\mathbb{R}_+^n$, or $|z| \rightarrow +\infty$

Since $z^* \gg 0$ it makes sense to define

$$\tilde{A} := A + diag({z^*}^{-1})\, B \tag{5.65}$$

$$\tilde{\tilde{A}} := \tilde{A} + diag\left[-\frac{b_1(z)}{z_1 z_1^*}, \ldots, -\frac{b_n(z)}{z_n z_n^*} \right] \tag{5.66}$$

and then the function $H : \mathbb{R}^{n*} \longrightarrow \mathbb{R}$

$$H(z) := (z - z^*)^T\, W \tilde{A}\, (z - z^*) - \sum_{i=1}^{n} w_i\, \frac{b_i(z)}{z_i z_i^*}\, (z_i - z_i^*)^2 \tag{5.67}$$

$$= (z - z^*)^T W\tilde{\tilde{A}} (z - z^*)$$

This function provides the derivative of V along the trajectories of the ODE system (5.63).

Suppose that either

a) for any $z \in \mathbb{R}^n$: $z^T W \tilde{A} z \leq 0$

or

b) for any $z \in \mathbb{R}^n$: $z^T W \tilde{\tilde{A}} z \leq 0$

Then, in either case

$$H(z) \leq 0 \,. \tag{5.68}$$

Let us now consider the parabolic system (5.58), subject to the boundary condition (5.58b).

Homogeneous Neumann boundary conditions allow (5.58) to keep z^* (constant with respect to both x and t) as an equilibrium solution.

If we take as initial condition a function $\overline{u} \in X$, $\overline{u} \gg 0$, and assume that f leaves $\mathbb{R}_+^{n*}$ invariant, we may consequently assume that, for any $t \geq 0$, $u(t) = U(t)\,\overline{u} \gg 0$, by comparison theorems [204; p.202] (see Appendix B, Section B.2.1).

It makes sense for $t \geq 0$ to define

$$y_i(x;t) := \ln \frac{u_i(x;t)}{z_i^*} \,, \qquad x \in \overline{\Omega} \tag{5.69}$$

and then the function

$$U : \overline{\Omega} \times \mathbb{R}_+ \longrightarrow \mathbb{R}_+$$

as follows

$$U(x;t) := V(u(x;t)) \,, \qquad (x,t) \in \overline{\Omega} \times \mathbb{R}_+ \,. \tag{5.70}$$

The derivative of U along the trajectories of system (5.58), (5.58b) is given by

$$\frac{\partial U}{\partial t} = grad\,V(u) \cdot \frac{\partial u}{\partial t}$$

$$= \sum_{i=1}^{n} w_i\,(u_i - z_i^*)\,\frac{\partial y_i}{\partial t}, \qquad \text{in } \overline{\Omega} \times (0, +\infty)\,;$$

while

$$\Delta U = \sum_{i=1}^{n} w_i \left\{ (u_i - z_i^*)\,\Delta y_i + u_i |grad\,y_i|^2 \right\}, \text{ in } \Omega \times (0, +\infty)$$

so that

$$\frac{\partial U}{\partial t} - \Delta U = \sum_{i=1}^{n} w_i \left\{ (u_i - z_i^*) \left[\frac{\partial y_i}{\partial t} - \Delta y_i \right] - u_i\,|grad\,y_i|^2 \right\}, \qquad \text{in } \Omega \times (0, +\infty)\,.$$

By the definition (5.69) and Eqn. (5.58) we get

$$\frac{\partial y_i}{\partial t} - \Delta y_i - |grad\,y_i|^2 = \frac{1}{u_i} \left[\frac{\partial u_i}{\partial t} - \Delta u_i \right]$$

$$= \frac{1}{u_i}\, f_i(u)\,.$$

Thus, for the particular choice (5.62)

(5.71) $$\frac{\partial U}{\partial t} - \Delta U = -\sum_{i=1}^{n} w_i z_i^*\,|grad\,y_i|^2 + H(u)$$

So, if we assume that a positive diagonal matrix $W = diag(w_1, \ldots, w_n)$ exists such that either a) or b) imply (5.68), we obtain from (5.71) the following

(5.72) $$\frac{\partial U}{\partial t} - \Delta U \leq 0\,.$$

At this point we may import most of the reasoning in [187] to prove the GAS of the positive equilibrium z^* of the ODE system (see Eqn. (5.64)) with respect to the reaction-diffusion system (5.58), (5.58b).

Since $\overline{u} \in C^2(\overline{\Omega})$, clearly it will bounded, so that also $0 \leq U(x;0) \leq \alpha$ (for some $\alpha > 0$). By the main comparison theorem (Appendix B, Section B.2.1) (5.72) will imply

$$0 \leq U(x;t) \leq \alpha \tag{5.73}$$

for the whole interval of existence of the solution of problem (5.58), (5.58b), (5.58o) with initial condition $\overline{u}(x)$.

By the definition of U, (5.73) implies an a priori estimate for u_i

$$0 < \gamma \leq u_i(x;t) \leq \delta$$

in the whole interval of existence of the solution u ; this implies on one side the global existence of $u(x;t),\ x \in \overline{\Omega},\ t \in \mathbb{R}_+$.

On the other hand it implies the boundedness of the trajectory $\Gamma_+(\overline{u})$ in X, hence its precompactness [218].

Further it can be shown [187]

Theorem 5.21. *For any $\overline{u} \in X$ (sufficiently smooth) $\omega(\overline{u})$ may contain only constants $c \in \mathbb{R}^n_+$.*

Due to the fact that the positive trajectory $\Gamma_+(\overline{u})$ is precompact, by Theorem 5.20 we know that the positive limit set $\omega(\overline{u})$ is nonempty, compact, connected, and invariant; moreover (5.61) holds, for $u^o = \overline{u}$.

To conclude our discussion we need to have more information about the asymptotic behavior of the solutions of the PDE problem (5.58).

Since $\omega(\overline{u})$ is made of constants, independent of $x \in \overline{\Omega}$, we may prove the following [187].

Lemma 5.22. *For any $\overline{u} \in X$ (sufficiently smooth) $\omega(\overline{u})$ is positively invariant with respect to the ODE system (5.63) (i.e. if $c \in \omega(\overline{u})$, then $\Gamma_+(c) \subset \omega(\overline{u})$, where $\Gamma_+(c)$ is the trajectory of the ODE system (5.63) with initial condition $c \in \mathbb{R}^n$), and $H(u)$ vanishes on $\omega(\overline{u})$.*

Proof. We already know that $\omega(\overline{u})$ is positively invariant for the PDE problem because of Theorem 5.20 . But, due the fact that the PDE system has zero Neumann boundary conditions, the solution $z(t;c)$ of the ODE problem (5.63) with initial condition $c \in \omega(\overline{u})$ will be a space independent solution of the PDE system itself, so that its trajectory $\Gamma_+(c)$ will all be contained in $\omega(\overline{u})$.

The proof of the second part of the lemma can be found in [187].

We are ready to state our main theorem.

Theorem 5.23. *The positive limit set $\omega(\overline{u})$ of the PDE system (5.58) with initial condition $\overline{u} \in X$, $\overline{u} \gg 0$ (sufficiently smooth) is contained in the largest invariant (with respect to the ODE system (5.63)) subset M of*

$$\mathcal{R} := \left\{ z \in \mathbb{R}^n_+ \,\middle|\, \dot{V}(z) = 0 \right\}.$$

Further, if $u(t)$ is the solution of (5.58) with initial condition $\overline{u}$, we have

$$\lim_{t \to +\infty} dist(u(t), M) = 0 \tag{5.74}$$

in $C^2\left(\overline{\Omega}\right)$.

Proof. We already know that for any $c \in \omega(\overline{u})$, $\Gamma_+(c) \subset \omega(\overline{u})$, with respect to the ODE system (5.63). Further $\dot{V}(c) = H(c) = 0$ so that

$$\omega(\overline{u}) \subset \mathcal{R}\,.$$

These two facts imply that $\omega(\overline{u})$ must be contained in the largest invariant (with respect to the ODE system (5.63)) subset of $\mathcal{R}$.

The above theorem relates the asymptotic behavior of the PDE system (5.58) to the behavior of the corresponding ODE system (5.63).
In particular in fact, if we know that M reduces to the equilibrium point z^* of the ODE system, then (5.74) implies that

$$\lim_{t \to +\infty} dist(U(t)\,\overline{u},\, z^*) = 0\,, \qquad \text{in } C^2\left(\overline{\Omega}\right) \tag{5.75}$$

for any initial condition $\overline{u} \in C^2\left(\overline{\Omega}\right)$, $\overline{u} \gg 0$, sufficiently smooth to guarantee that the solution $u(t) := U(t)\,\overline{u}$ is a classical solution of the PDE problem (5.58), (5.58b), (5.58o).
As far as stability is concerned, suppose we start with an initial condition $\overline{u} \in C^2\left(\overline{\Omega}\right)$, $\overline{u} \gg 0$ such that for some δ

$$|\overline{u}(x) - z^*| < \delta$$

then a $\beta > 0$ exists such that

$$U(x;0) := V(\overline{u}(x)) \leq \beta \, .$$

Since we have proven that $U(x;t) := V(u(t,x))$ satisfies (5.72) as a consequence we have, by (5.73), that

$$0 \leq U(x;t) \leq \beta \qquad \text{for any } t \geq 0$$

and this will imply that for some $\gamma > 0$

$$|u(x;t) - z^*| \leq \gamma \qquad \text{in } \overline{\Omega} \times (0, +\infty)$$

and this shows that z^* is stable for the PDE system.

This, together with (5.75), gives the GAS of z^* for the PDE system in $\overset{\circ}{C}_+ \left(\overline{\Omega}\right)$.

Remark. We make the final remark that conditions for M to reduce to the only equilibrium point z^* of the ODE problem have already been given in Sect. 2, Lemma 2.2, based on graph theoretical methods.

Conditions (A) and (B) in Sect. 2 are just sufficient conditions for a) and b) to hold, respectively.

Thus Theorem 5.23 may be seen as a direct extension of our results for GAS of the spatially homogeneous epidemic systems analyzed in Sect. 2 to the corresponding spatially heterogeneous case with linear diffusion .

6. Age structure

In this section we introduce a dependence of the force of infection upon the chronological age of individuals participating in the epidemic.

Age has been recognized as an important factor in the dynamics of epidemic processes since 1760 by Bernoulli [32] when reporting the data of a smallpox epidemic .

Age dependent models have been analyzed in [70, 121], and great attention has been paid in connection with the analysis of real epidemics by Dietz and Schenzle [84].

From the mathematical point of view a good reference is the monograph [168].

A stochastic version of age-dependent epidemic systems has been proposed more recently in [47] which allows, by means of martingale theory for point processes, the statistical analysis of age-dependent epidemic data.

Here we shall report about the results obtained by Busenberg et al. [41] about the existence and stability of nontrivial endemic states for SIS epidemic systems with constant total population. This is also to show the mathematical methods employed for the analysis of age-dependent epidemic systems.

The total population has an age structure expressed in terms of its age density $p(a;t)$, $a \in \mathbb{R}_+, t \in \mathbb{R}_+$. It is divided into two classes; the susceptible population with age density $s(a;t)$ and the infective population with age density $i(a;t)$, so that

$$p(a;t) = s(a;t) + i(a;t), \qquad a,t \in \mathbb{R}_+. \tag{6.1}$$

Each individual is subject to an age-dependent death rate $\mu(a)$. New individuals are produced at an age-dependent birth rate $\beta(a)$.

The deterministic mathematical model currently accepted for the evolution of the age-dependent populations is based on the so called McKendrick-van Foerster equation [167, 214]

$$\frac{\partial}{\partial t} p(a;t) + \frac{\partial}{\partial a} p(a;t) + \mu(a)\, p(a;t) = 0 \tag{6.2}$$

for $a, t \in \mathbb{R}_+$, subject to the boundary condition

$$p(0;t) = \int_0^\infty \beta(a) p(a;t)\, da, \qquad t \geq 0 \tag{6.3}$$

and to the initial condition

$$p(a;0) = p^o(a), \qquad a \geq 0 \tag{6.4}$$

where $p^o(a), a \in \mathbb{R}_+$ is the initial age distribution of the total population.

Typical assumptions on the parameters are

(H1) β, μ, p^o are nonnegative, piecewise continuous functions on $[0, \infty)$.

(H2) $\beta(a) > 0$ for $a \in (A_o, A)$
$\beta(a) = 0$ for $a \notin (A_o, A)$

(H3) $\displaystyle\int_0^\infty \exp\left(-\int_0^a \mu(\sigma)\, d\sigma\right) da < \infty$.

Note that (H3) implies that $\displaystyle\int_0^\infty \mu(\sigma)\, d\sigma = \infty$.

Under the above assumptions, system (6.2)-(6.4) has a steady state solution (see e.g. [70, 121]) given by

$$p_\infty(a) := b_o \exp\left(-\int_0^a \mu(\sigma)\, d\sigma\right) , \quad a \geq 0 , \tag{6.5}$$

iff the net population reproduction rate R equals 1:

$$R := \int_0^\infty \beta(a) \exp\left(-\int_0^a \mu(\sigma)\, d\sigma\right) da = 1 . \tag{6.6}$$

($b_o \geq 0$ is an arbitrary parameter related to the total population size).

By referring to the analysis carried out in [41] we shall assume that (6.6) holds and the total population has reached its steady state age-distribution $p_\infty(a)$, $a \in \mathbb{R}_+$.

In such a population we introduce an SIS epidemic.

6.1. An SIS model with age structure

According to the basic definitions in Sect.2 the total population $\{p_\infty(a),\ a \in \mathbb{R}_+\}$ is divided into two subgroups; the susceptible population, with age distribution $\{s(a;t),\ a \in \mathbb{R}_+\}$, and the infective population, with age distribution $\{i(a;t),\ a \in \mathbb{R}_+\}$, at time $t \geq 0$, so that

$$(6.7) \qquad p_\infty(a) = s(a;t) + i(a;t) \ , \qquad a \geq 0 \ , \quad t \geq 0 \ .$$

Given (6.7), the epidemic process can be described by the evolution of the infective population. The basic equations are given in [167] for an early reference; a more detailed discussion can be found in [198]. In accordance with our approach (see Sect.1 and Sect.5.1), as for the space structure the force of infection acting on a susceptible individual of age a, $g(i(\cdot;t))(a)$ depends a priori on the overall distribution of the infective population at time t, $i(\cdot;t)$.

A possible choice, suggested by Schenzle [198] is the following

$$(6.8) \qquad g(i(\cdot;t))(a) = \int_0^\infty k(a,a')\, i(a';t)\, da'$$

where $k(a,a')$ describes the action of the infectives of age $a' \in \mathbb{R}_+$ on the susceptibles of age $a \in \mathbb{R}_+$.

As a consequence the infection process is described by

$$(6.9) \qquad g(i(\cdot;t))(a)\, s(a;t) \ .$$

For the case $k(a,a') = \delta(a-a')\, k_o(a)$ (δ is the Dirac function) we get [40.1]

$$(6.10) \qquad g(i(\cdot;t))(a) = k_o(a)\, i(a;t)$$

The case (6.10) is known as "intracohort" infection process; while (6.8) is better known as "intercohort" infection process.

We may wish to remind here that the case $k(a,a') = k$, constant, has been investigated in [80, 98]; $k(a,a') = k(a)$, in [101, 220]; $k(a,a') = k_1(a)\, k_2(a')$, in [84, 99].

If we consider an SIS model, the evolution of the age distribution $\{i(a;t),\ a \in \mathbb{R}_+\}$, $t \in \mathbb{R}_+$, is described by

$$(6.11) \quad \frac{\partial}{\partial t} i(a;t) + \frac{\partial}{\partial a} i(a;t) + \mu(a)\, i(a;t) = g(i(\cdot\,;t))(a)\, s(a;t) - \gamma(a)\, i(a;t)$$

$$(6.12) \qquad i(0;t) = q \int_0^\infty \beta(a) i(a;t)\, da$$

$$(6.13) \qquad i(a;0) = i^o(a)$$

where $\gamma(a)$, $a \in \mathbb{R}_+$ is the recovery rate of infectives (going back into the susceptible class).

The parameter $0 \leq q \leq 1$ in (6.12) is the probability for the disease to be vertically transmitted from infective parents to newborns. When $q = 0$ there is no vertical transmission and hence condition (6.12) becomes

$$i(0;t) = 0 \tag{6.12'}$$

that is all newborns are susceptible.

We shall report here the analytical results obtained in [41] about the threshold theorems.

6.1.1. The intracohort case

If we choose the case (6.10) equation (6.11) becomes

$$\frac{\partial}{\partial t} i(a;t) + \frac{\partial}{\partial a} i(a;t) + \mu(a)\, i(a;t) = k_o(a)\, i(a;t)(p_\infty(a) - i(a;t)) - \gamma(a)\, i(a;t) \tag{6.14}$$

where we have taken (6.7) into account.

We shall make the further assumption that

(H4) γ and k_o are nonnegative piecewise continuous functions on $[0, +\infty)$ and k_o is bounded.

System (6.14), (6.12), (6.13) can be explicitly solved along the characteristic lines $t - a = const$, and we obtain the following :

$$i(a;t) = \begin{cases} i_1(a;t)\,, & \text{if } \ a \geq t \\ i_2(a;t)\,, & \text{if } \ a < t \end{cases} \tag{6.15}$$

where

$$i_1(a;t) = \frac{i^o(a-t)\exp\left(\int_0^t \alpha(a-t+\sigma)\, d\sigma\right)}{1 + i^o(a-t)\int_0^t \exp\left(\int_0^\tau \alpha(a-t+\sigma)\, d\sigma\right) k_o(a-t+\tau)\, d\tau}$$

$$i_2(a;t) = \frac{i(0,t-a)\exp\left(\int_0^a \alpha(\sigma)\, d\sigma\right)}{1 + i(0,t-a)\int_0^a \exp\left(\int_0^\tau \alpha(\sigma)\, d\sigma\right) k_o(\tau)\, d\tau}$$

and

$$\alpha(\sigma) = -\mu(\sigma) - \gamma(\sigma) + k_o(\sigma)p_\infty(\sigma), \qquad \sigma \geq 0 \tag{6.16}$$

To completely explicit $i(a;t)$ we need to take into account the boundary condition (6.12).

6.1.1.1. The case $q = 0$

For the case $q = 0$ (no vertical transmission), we apply (6.12') to get

$$i(a;t) = \begin{cases} i_1(a;t), & \text{if } \ a \geq t \\ 0 & \text{if } \ a < t \end{cases} \tag{6.17}$$

where $i_1(a;t)$ is defined in the Eqn. (6.15) and following.

If we denote, for $a \geq 0$,

$$\begin{aligned} E(a) &:= \exp\left(\int_0^a \alpha(\sigma)\, d\sigma\right) \\ &= \frac{p_\infty(a)}{b_o} \exp\left(\int_0^a [-\gamma(\sigma) + k_o(\sigma)p_\infty(\sigma)]\, d\sigma\right) \end{aligned} \tag{6.18}$$

from (6.17) we get, for the cohort $a_o = a - t \geq 0$

$$i(a_o + t; t) = \frac{i^o(a_o)\dfrac{E(a_o+t)}{E(a_o)}}{1 + i^o(a_o)\displaystyle\int_o^t \left(\frac{E(a_o+\tau)}{E(a_o)}\right) k_o(a_o+\tau)\, d\tau}, \qquad t \geq 0 \tag{6.19}$$

Since

$$\lim_{a\to\infty} E(a) = 0 \quad ; \quad \int_0^\infty E(a)\, da < +\infty \tag{6.20}$$

from (6.19) we get, for any cohort $a_o \geq 0$,

$$\lim_{t\to\infty} i(a_o + t; t) = 0 \ . \tag{6.21}$$

That is the cohort of infectives which has age a_o at time $t = 0$ eventually vanishes. We may then conclude that when there is no vertical transmission of the disease, any epidemic dies off due to the aging process.

6.1.1.2. The case $q > 0$

The case $q > 0$ requires a more elaborate treatment based on an integral equation for the quantity

$$u(t) := i(0;t)\,. \tag{6.22}$$

By substituting (6.15) into (6.12) we obtain for $u(t)$ the integral equation

$$u(t) = F(t) + \int_0^t G(a, u(t-a))\,da\,, \qquad t \geq 0 \tag{6.23}$$

where

$$F(t) := \int_t^\infty \frac{q\beta(a)E(a)\,i^o(a-t)}{E(a-t) + i^o(a-t)\displaystyle\int_{a-t}^a E(\tau)k_o(\tau)\,d\tau}\,da\;, \quad t \geq 0 \tag{6.24}$$

$$\begin{aligned} G(a,z) &:= \frac{q\beta(a)E(a)z}{1 + z\displaystyle\int_0^a E(\tau)k_o(\tau)\,d\tau}\;, \qquad a \geq 0, \quad z \geq 0 \\ &= D(a,z)\,z \end{aligned} \tag{6.25}$$

where we have denoted by

$$D(a,z) := \frac{q\beta(a)E(a)}{1 + z\int_0^a E(\tau)k_o(\tau)\,d\tau}\;, \qquad a \geq 0, \quad z \geq 0\,. \tag{6.26}$$

If we make the assumptions (H1)-(H4) we get

$$F(t) = 0 \qquad \text{for} \quad t > A \tag{6.27}$$

$$\begin{aligned} G(a,z) &> 0 \qquad \text{for} \quad a \in (A_o, A) \\ G(a,z) &= 0 \qquad \text{for} \quad a \notin (A_o, A) \end{aligned} \tag{6.28}$$

Furthermore for any $a \in (A_o, A)$:

$$z \to G(a,z) \qquad \text{is an increasing function} \tag{6.29}$$

$$z \to D(a,z) \qquad \text{is a decreasing function.} \tag{6.30}$$

Because of (6.27) and (6.28) the asymptotic behavior of equation (6.23) is the following

$$v(t) = \int_0^A G(a, v(t-a))\,da\,, \qquad t \to +\infty \tag{6.31}$$

so that, if we are interested in constant solutions of (6.31):

$$V = \int_0^A G(a,V)\,da \tag{6.32}$$

or better

$$V = V \int_0^A D(a,V)\,da\,. \tag{6.33}$$

Thus, either $V = 0$, or $V \neq 0$ as given by

$$1 = \int_0^A D(a,V)\,da \tag{6.34}$$

Since the function

$$\Delta(V) = \int_0^A D(a,V)\,da$$

is decreasing with limit zero as $V \to \infty$,equation (6.34) will have one and only one solution, iff the threshold condition

$$T := \Delta(0) > 1 \tag{6.35}$$

is satisfied.

The threshold parameter

$$\begin{aligned} T &= \int_0^A q\beta(a)E(a)\,da \\ &= \frac{q}{b_o}\int_0^A \beta(a)p_\infty(a)\exp\left(\int_0^a [-\gamma(\sigma)+k_o(\sigma)p_\infty(\sigma)]\,d\sigma\right)da \end{aligned} \tag{6.36}$$

can be interpreted as a net infection-reproduction rate (number).

By summarizing the above results we get

Theorem 6.1. *Let the "threshold parameter" T be defined as in (6.36).*

(a) If $T \leq 1$, then equation (6.32) has only the trivial constant solution $V \equiv 0$.

b) If $T > 1$, then it also admits a nontrivial solution $V_\infty > 0$ which is obtained as a solution of (6.34).

As far as the stability of the constant solutions 0 and V_∞ is concerned, the following main theorem is shown in [41] by monotone iteraction techniques.

Theorem 6.2. *Let T be defined as before.*

a) If $T \leq 1$ *, then* $\lim_{t\to\infty} u(t) = 0$.

b) If $T > 1$ *, then* $\lim_{t\to\infty} u(t) = V_\infty$.

6.1.2. The intercohort case

The results obtained in Sect. 6.1 for the intracohort case can be extended to the general case in which the force of infection is

$$g(i(\cdot\,;t))\,(a) = k_o(a)\,i(a;t) + \int_0^\infty k(a,a')\,i(a';t)\,da' \tag{6.37}$$

The pure intercohort case ($k_o = 0$) had been analyzed in [41] obtaining only partial local stability results. More recently in [42] a final answer has been given to the above problem by using semigroup theoretic methods in Banach spaces. It is interesting to report about this case here, since, as expected from the analysis in Sect.2.3.1 and Sect. 4.3 , SIS models belong to Class B and show a "quasi-monotone" behavior that implies a monotone evolution operator for which periodic solutions are ruled out (see e.g. [119]), and a nontrivial steady state, whenever it exists, is globally asymptotically stable. Thus showing that quasimonotone systems are "robust" also with respect to age structure.

In this case, with the force of infection (6.37), and $q = 0$, system (6.11)-(6.13) becomes

$$\frac{\partial}{\partial t}u(a,t) + \frac{\partial}{\partial a}u(a,t) + \mu(a)u(a,t) \tag{6.38}$$

$$= \tilde{g}(u(\cdot,t))(a)\,(1 - u(a,t)) - \gamma(a)u(a,t)$$

$$u(0,t) = 0 \tag{6.39}$$

$$u(a,0) = u^o(a) \tag{6.40}$$

for $a \geq 0$ and $t \geq 0$, where we have introduced the adimensional fraction of infectives

$$u(a,t) = \frac{i(a;t)}{p_\infty(a)} \tag{6.41}$$

and

$$\tilde{g}(u(\cdot,t))(a) = k_o(a)p_\infty(a)u(a,t) + \int_0^\infty k(a,a')p_\infty(a')u(a',t)\,da' \tag{6.42}$$

In [42] it is assumed that a maximum age $a^\dagger$ exists such that

(K1) $\beta(\cdot)$ is non-negative and belongs to $L^\infty(0,a^\dagger)$

(K2) $\mu(\cdot)$ is non-negative and measurable

(K3) $R = \int_0^{a^\dagger} \beta(a) e^{-\int_0^a \mu(a')\, da'}\, da = 1$

so that we can assume that the total population has reached its steady state $\{p_\infty(a), a \in \mathbb{R}_+\}$ given by (6.5).

Further it is assumed that

(K4) $\gamma(\cdot)$ and $k_o(\cdot)$ are non-negative and belong to $L^\infty(0, a^\dagger)$

(K5) $k(a, a'),\quad a, a' \in [0, a^\dagger]$, is measurable and there exists a positive constant $\epsilon > 0$ and two non-negative functions $k_1, k_2 \in L^\infty(0, a^\dagger)$ such that

α) $\epsilon k_1(a) k_2(a') \le k(a, a') p_\infty(a') \le k_1(a) k_2(a'),\quad$ in $\quad [0, a^\dagger]$.

β) there exist $\quad 0 \le a_1, a_2, b_1, b_2 \le a^\dagger \quad$ such that

$$a_1 < b_1, \quad a_2 < b_2, \quad a_1 < b_2$$

and

$$k_1(a) > 0, \qquad \text{if} \quad a_1 < a < b_1$$

$$k_2(a) > 0, \qquad \text{if} \quad a_2 < a < b_2\ .$$

The mathematical set up which has been used to analyze this problem refers to the Banach space $L^1(0, a^\dagger)$ so that the solutions $\{u(\cdot, t), t \in \mathbb{R}_+\}$ of problem (6.38)-(6.40) are looked for as elements of the closed convex set

$$C = \{f \in L^1(0, a^\dagger) \mid 0 \le f(a) \le 1 \quad \text{a.e.}\} \tag{6.43}$$

The following theorem has been proven [42].

Theorem 6.3. *Let the initial condition $u^o \in C$. Then problem (6.38)- (6.40) has a unique mild solution [179] in C. This defines a flow $\{S(t)u^o, t \in \mathbb{R}_+\}$ [119] which has the following properties*

$$S(t)C \subset C \tag{6.44}$$

$$\text{if} \quad u^o \le v^o \quad \text{then} \quad S(t)u^o \le S(t)v^o \tag{6.45}$$

$$\text{if} \quad 0 \le \xi \le 1 \quad \text{then} \quad \xi S(t)u^o \le S(t)(\xi u^o) \tag{6.46}$$

We wish to point out that a mild solution of system (6.38)-(6.40) is essentially a solution of a suitable corresponding integral formulation of the same problem [26], so that it is not required that $\{S(t)u^o, t \in \mathbb{R}_+\}$ admits time derivatives (for a more detailed discussion we refer to [42] and [179]; see also [168] and [220]).

Condition (6.45) gives the monotonicity of the evolution operator $\{S(t), t \in \mathbb{R}_+\}$, while (6.46) expresses its sublinearity.

An endemic solution u_∞ of system (6.38)-(6.40) is a fixed point of the evolution operator $\{S(t), t \in \mathbb{R}_+\}$:

$$S(t)u_\infty = u_\infty\,, \qquad t \in \mathbb{R}_+ \;. \tag{6.47}$$

Unfortunately the threshold theorem as given in [42] refers to an abstract formulation of system (6.38)-(6.40).

Anyway a parameter ρ , the spectral radius of a suitable operator defined on the Banach space $L^1(0, a^\dagger)$, is introduced so that

Theorem 6.4.

a) *If* $\rho \leq 1$ *then no nontrivial endemic states exist for system (6.38)-(6.40).*

b) *if* $\rho > 1$ *then a unique nontrivial endemic state* u_∞ *exists for system (6.38)-(6.40), in addition to the trivial one.*

The stability properties of the endemic equilibria are established by the following

Theorem 6.5. *Assume no nontrivial endemic equilibrium exists, then for any initial condition* $u^o \in C$ *we have*

$$\lim_{t\to\infty} S(t)u^o = 0 \;.$$

To consider the nontrivial case, we introduce the concept of nontrivial initial condition, i.e. a $u^o \in C$ such that

$$\int_t^{a^\dagger} k_2(a)\, u^o(a-t)\, da > 0, \qquad \text{for some} \quad t \geq 0 \;.$$

Theorem 6.6. *Let* u_∞ *be the unique nontrivial equilibrium. Then for any nontrivial initial condition* u^o *we have*

$$\lim_{t\to\infty} S(t)u^o = u_\infty \;.$$

If u^o *is not nontrivial then*

$$S(t)\, u^o = 0\,, \qquad \text{for any} \quad t \geq a^\dagger$$

.

The monotonicity of the evolution operators $\{S(t), t \in \mathbb{R}_+\}$ implies the GAS of the trivial endemic state in Theorem 6.5 and of the nontrivial endemic state in Theorem 6.6.

Further extension to the case of vertically transmitted diseases is considered in [43].

6.2. An SIR model with age structure [123]

In this section we report the results obtained recently by Inaba [123] about the existence and stability of an SIR model with age structure as formulated by Greenhalgh [99].

The total population has an age structure expressed in terms of its age density $p(a;t)$, $a \in [0, a^\dagger]$, $t \geq 0$ (it is assumed a maximum demographic age $a^\dagger > 0$).

It is divided into three classes; the susceptible class, with age density $s(a;t)$; the infective class with age density $i(a;t)$, and the removed (immune) class, with age density $r(a;t)$, so that

$$(6.48) \qquad p(a;t) = s(a;t) + i(a;t) + r(a;t), \quad a \in [0, a^\dagger], \ t \geq 0.$$

As in Sect. 6.1, each individual is subject to an age-dependent death rate $\mu(a)$; new individuals are produced at an age dependent birth rate $\beta(a)$, $0 \leq a \leq a^\dagger$.

The evolution of the total density $p(a;t)$ is then governed by system (6.2) for $a \in [0, a^\dagger]$, $t \geq 0$, subject to the boundary condition

$$(6.49) \qquad p(0;t) = \int_0^{a^\dagger} \beta(a)\, p(a;t)\, da \ , \qquad t \geq 0$$

and to an initial condition (6.4).

Assumptions (H1)-(H2) are kept but (H3) now is

$$(H3') \qquad \int_0^{a^\dagger} \exp\left(-\int_0^a \mu(\sigma)\, d\sigma\right) da < +\infty$$

and that

$$(6.50) \qquad \int_0^{a^\dagger} \mu(\sigma)\, d\sigma = +\infty \ .$$

System (6.2), (6.49), (6.4) has a steady state solution given by

$$(6.51) \qquad p_\infty(a) = b_o \exp\left(-\int_0^a \mu(\sigma)\, d\sigma\right), \qquad a \in [0, a^\dagger]$$

iff the net population reproduction rate R equals one,

$$(6.52) \qquad R = \int_0^{a^\dagger} \beta(a) \exp\left(-\int_0^a \mu(\sigma)\, d\sigma\right) da = 1 \ .$$

We shall assume, as in Sect. 6.1, that the total population has reached its stationary demographic state (6.51).

In such a population we introduce an SIR epidemic, so that (6.48) has to be rewritten as

$$p_\infty(a) = s(a;t) + i(a;t) + r(a;t) \ , \quad a \in [0, a^\dagger] \, , \ t \geq 0 \, . \tag{6.53}$$

If we consider for the force of infection the same form as in (6.8), and a removal rate $\gamma > 0$, which now is assumed age independent, the evolution equations for the epidemic system are

$$\left(\frac{\partial}{\partial t} + \frac{\partial}{\partial a}\right) s(a;t) = -g(i(\cdot,t))(a)\, s(a;t) - \mu(a)\, s(a;t) \tag{6.54a}$$

$$\left(\frac{\partial}{\partial t} + \frac{\partial}{\partial a}\right) i(a;t) = -g(i(\cdot,t))(a)\, s(a;t) - (\mu(a) + \gamma)\, i(a;t) \tag{6.54b}$$

$$\left(\frac{\partial}{\partial t} + \frac{\partial}{\partial a}\right) r(a;t) = \gamma\, i(a;t) - \mu(a)\, r(a;t) \tag{6.54c}$$

for $a \in \left]0, a^\dagger\right[\, , \ t > 0$, subject to the boundary conditions

$$s(0;t) = \int_0^{a^\dagger} \beta(a)\, p(a;t)\, da \tag{6.55a}$$

$$i(0;t) = 0 \tag{6.55b}$$

$$r(0;t) = 0 \tag{6.55c}$$

Under the assumption of demographic equilibrium

$$\int_0^{a^\dagger} \beta(a)\, p(a;t)\, da = \int_0^{a^\dagger} \beta(a)\, p_\infty(a)\, da =$$
$$= b_o \int_0^{a^\dagger} \beta(a) \exp\left(-\int_0^a \mu(\sigma)\, d\sigma\right) da = b_o R = b_o \, ,$$

so that the boundary condition (6.55a) can be rewritten as

$$s(0;t) = b_o \, . \tag{6.55a'}$$

We may renormalize all densities so that

$$s(a;t) + i(a;t) + r(a;t) = 1 \, , \qquad a \in [0, a^\dagger] \, , \ t \geq 0 \ , \tag{6.56}$$

in which case

$$(6.55a'') \qquad s(0;t) = 1\,, \qquad t \geq 0\,.$$

The force of infection, under condition (6.56) will be given by

$$(6.57) \qquad g(i(\cdot;t))(a) = \int_0^{a^\dagger} k(a,a')\,p_\infty(a')\,i(a';t)\,da'$$

with $p_\infty(a)$ given by (6.51).

System (6.54), (6.55) has to be supplemented by suitable initial conditions

$$(6.58) \quad s(a;0) = s^o(a); \quad i(a;0) = i^o(a); \quad r(a;0) = r^o(a), \qquad a \in \left[0, a^\dagger\right]\,.$$

We shall assume in the sequel that

$$(H4') \qquad k(\cdot,\cdot) \in L^\infty\left(\left[0,a^\dagger\right] \times \left[0,a^\dagger\right]\right)\,.$$

We note that due to condition (6.56) it suffices to analyze system (6.54a), (6.54b) with the boundary conditions (6.55a"), (6.55b) subject to the initial conditions (6.58).

Under these assumptions (H1), (H2), (H3'), (H4') it is possible to show, by semigroup theoretical methods that this problem is well posed in the Banach space $X := L^1\left(0,a^\dagger;\, \mathbb{R}^2_+\right)$.

In fact, if we define

$$(6.59) \qquad \Omega_o := \left\{f = (f_1,f_2)^T \in X \mid 0 \leq f_i \leq 1,\ \text{a.e. } i = 1,2\right\}$$

the following theorem holds [123]

Theorem 6.7. *Let the initial condition $u^o := (s^o(\cdot), i^o(\cdot)) \in \Omega_o$. Then problem (6.54 a,b), (6.55 a",b), (6.58) has a unique mild solution [26, 179] in Ω_o. If we further assume that the initial conditions are absolutely continuous in $\left[0,a^\dagger\right]$, then the initial value problem admits a unique global classical solution in Ω_o.*

Let us now look for steady states $u^* := (s^*(\cdot), i^*(\cdot))^T$ of our system. It is not difficult to show that it must satisfy the following

$$(6.60a) \qquad s^*(a) = \exp\left(-\int_0^a g(i^*)(\sigma)\,d\sigma\right)$$

$$(6.60b) \quad i^*(a) = \int_0^a \exp(-\gamma(a-\sigma))\,g(i^*)(\sigma)\,\exp\left(-\int_0^\sigma g(i^*)(\eta)\,d\eta\right)\,d\sigma$$

$$(6.60c) \qquad \lambda^*(a) := g(i^*)(a) = \int_0^{a^\dagger} k(a,\sigma)\,p_\infty(\sigma)\,i^*(\sigma)\,d\sigma\,.$$

From (6.60b) and (6.60c) we obtain an equation for $\lambda^*(a)$:

$$\lambda^*(a) = \int_0^{a^\dagger} \phi(a,\sigma)\,\lambda^*(\sigma)\,\exp\Big(\Big(-\int_0^{\sigma}\lambda^*(\eta)\,d\eta\Big)\,d\sigma\ , \tag{6.61}$$

where

$$\phi(a,\sigma) := \int_\sigma^{a^\dagger} k(a,\tau)\,p_\infty(\tau)\,\exp(-\gamma(\tau-\sigma))\,d\tau\ .$$

It is clear that (6.61) always admits the trivial solution $\lambda^*(a) \equiv 0$ in $[0,a^\dagger]$, which corresponds to a disease-free steady state $u^* := (1,0)^T$. So we look for nontrivial solutions of (6.61).

If we denote by Φ the nonlinear operator

$$x \in E \longrightarrow \Phi(x)(a) := \int_0^{a^\dagger} \phi(a,\sigma)\,x(\sigma)\,\exp\left(-\int_0^{\sigma} x(\eta)\,d\eta\right)\,d\sigma \tag{6.62}$$

with $a \in [0,a^\dagger]$, and defined in the Banach space $E := L^1(0,a^\dagger)$, equation (6.61) is equivalent to a fixed point problem for Φ:

$$\lambda^* = \Phi(\lambda^*)$$

(Note that $L^\infty(0,a^\dagger) \subset E$).

The nonlinear operator Φ has a linear majorant $T : E \longrightarrow E$ defined by

$$x \in E \longrightarrow T(x)(a) := \int_0^{a^\dagger} \phi(a,\sigma)\,x(\sigma)\,d\sigma\ , \qquad a \in [0,a^\dagger] \tag{6.63}$$

which is positive with respect to the cone $E_+ := \{x \in E \,|\, 0 \le x\}$, i.e.

$$T(E_+) \subset E_+\ .$$

In addition to (H4') let us further assume that

(H5) a) $\lim_{h\to 0} \int_0^{a^\dagger} |k(a+h,\sigma) - k(a,\sigma)|\,da = 0$ uniformly in $\sigma \in \mathbb{R}$.
(k is extended by $k(a,\sigma) = 0$ for $a,\sigma \in (-\infty,0) \cup (a^\dagger,+\infty)$).

b) There exist numbers $\alpha > 0$ and $\epsilon > 0$ with $0 < \alpha < a^\dagger$, such that $k(a,\sigma) \ge \epsilon$ for a.e. $(a,\sigma) \in (0,a^\dagger) \times (a^\dagger - \alpha, a^\dagger)$.

Under the above assumptions [123] (see [160] for the terminology):

Lemma 6.8. *The linear operator T is nonsupporting and compact.*

As a consequence [160] the spectral radius $r(T)$ of the operator T is a positive eigenvalue of T, and it is the only positive eigenvalue with a positive eigenvector.

The following threshold theorem holds [123].

Theorem 6.9. *Let $r(T)$ be the spectral radius of the operator T defined in (6.63).*

a) *If $r(T) \le 1$, then the only nonnegative fixed point x of Φ is the trivial solution $x \equiv 0$.*
b) *If $r(T) > 1$, then Φ has at least one nontrivial positive solution.*

In order to have uniqueness of the nontrivial solution, a further assumption is made.

(H6) For all $(a, \sigma) \in [0, a^\dagger] \times [0, a^\dagger]$, the following inequality holds:

$$k(a,\sigma)\, p_\infty(\sigma) - \gamma \int_\sigma^{a^\dagger} k(a,\tau)\, p_\infty(\tau)\, \exp(-\gamma(\tau - \sigma))\, d\tau \ge 0\,.$$

Now observe that, from the definition of Φ, it follows that

$$\begin{aligned}\Phi(x)\,(a) &= \int_0^{a^\dagger} \phi(a,\sigma) \left(-\frac{d}{d\sigma}\right) \exp\left(-\int_0^\sigma x(\eta)\, d\eta\right) d\sigma \\ &= \phi(a,0) - \int_0^{a^\dagger} [k(a,\sigma)\, p_\infty(\sigma) - \gamma\phi(a,\sigma)]\, \exp\left(-\int_0^\sigma x(\eta)\, d\eta\right) d\sigma\end{aligned}$$

so that Φ is monotone increasing with respect to the cone E_+ .

Furthermore it can be shown [123], under the same assumption, that Φ is strongly concave in the sense of Krasnoselskii (see Appendix A, Section A.4.2).

As a consequence of positivity, monotonicity and strong concavity we know that (Theorem A.36 in Appendix A) the operator Φ has at most one positive fixed point. Thus

Theorem 6.10. *Under the assumption (H6), if $r(T) > 1$, then Φ admits a unique nontrivial positive solution.*

Remark. A sufficient condition for assumption (H6) to be verified is

(H6′) $\qquad k(a,\sigma)\, p_\infty(\sigma)$ is nonincreasing with respect to σ

.

A particular case of (H6') is $k(a,\sigma) \equiv k(a)$ independent of the age σ of the infectives, since $p_\infty(\sigma)$ is a decreasing function of σ.

Remark. Independently of the fact that (H6) holds, it can be easily shown (see e.g. [99]) that if $k(a,\sigma) = k_1(a)\,k_2(\sigma)$ (proportionate mixing assumption [84]), then there exists a unique nontrivial steady state under the condition

$$r(T) = \int_0^{a^\dagger} \phi(\sigma,\sigma)\,d\sigma \;> 1$$

.

In this case $f(\cdot)$ is the eigenvector of the operator T corresponding to the spectral radius $r(T)$.

We are left now with the problem of stability of the steady states.

Since the analysis in [123] contains technicalities which go beyond the interests of this presentation, we shall limit ourselves to reporting the main results.

Theorem 6.11.

a) *If $r(T) < 1$, then the trivial equilibrium is GAS for system (6.54), (6.55) with respect to positive initial conditions .*

b) *If $r(T) > 1$, then the trivial solution of system (6.54), (6.55) is unstable. If furthermore the nontrivial endemic state is such that*

(H7) $$i^*(a^\dagger) < e^{-\gamma a^\dagger}$$

then it is LAS.

Condition (H7), being an a priori condition on the steady state, is not very convenient in the applications.

In [123] it is shown that (H7) holds when $k(a,\sigma) \equiv k$, a constant.

We conclude this section, by remarking that again for SIR models with structures the problem of GAS is more difficult and in fact it left open for many relevant cases.

7. Optimization problems

7.1. Optimal control

The problem of controlling an epidemic may be stated as a typical optimal control problem whenever the relevant cost functions can be explicitly given.

This is a difficult task in general but anyway decision makers should pay sufficient attention to identifying such cost functions.

In the literature attempts in this direction can be found [102, 117].

Here we shall outline as an example the approach which has been followed in connection with the problem of controlling a specific class of man-environment-man epidemics, as described in Section 5.5 .

In Sect. 5.5 we have been discussing a threshold theorem for the existence of a nontrivial endemic state. In particular for a specific boundary condition ($\alpha = 0$ in (5.45)) the threshold parameter is given by

$$\tilde{\theta} = \frac{g'(0)}{a_{11}a_{22}} \sup_{x \in \Omega} \int_{\partial\Omega} k(x, x')\, d\sigma(x') \tag{7.1}$$

in the sense that if $\tilde{\theta} < 1$ no nontrivial endemic state may exist for the epidemic system (5.44), (5.45).

A natural question arises about the possibility of reducing the "size" of the epidemic by acting on the feedback kernel $k(x, x')$ by means of sanitation programs of the sewage system (see also [76]).

Certainly this implies a cost that has to be compared with the cost of the epidemic itself.

Suppose we modify $k(x, x')$ by a multiplication parameter $w(x, t)$ so that the new feedback operator becomes

$$H_w[u_2(\cdot, t)](x) = w(x, t) \int_{\Omega} k(x, x') u_2(x', t)\, dx' \tag{7.2}$$

at any time $t > 0$, and at any point $x \in \partial\Omega$.

During an observation time $T > 0$ the total cost of the sanitation program will be given by

$$J_1 = \int_0^T \int_{\partial\Omega} h(w(t, \sigma))\, d\sigma\, dt \ ; \tag{7.3}$$

h being the cost function of the sanitation program.

The cost J_1 has to be opposed to the total cost of the epidemic which is given by

(7.4) $$J_2 = \int_0^T \int_\Omega f(u_2(x,t))\,dx\,dt \ ,$$

f being the cost function of the epidemic.

Usually a penalty, or cost, is also associated with still having an infected human population at the end of the period of sanitation; an additional cost function may be added of the form

(7.5) $$J_3 = \int_\Omega l(u_2(x,T))\,dx$$

Typically the optimal control problem is thus the following :

Problem (P). *Minimize*

$$J(u_1, u_2, w) := J_1 + J_2 + J_3$$

on all (u_1, u_2, w) *subject to the "state" system*

(7.6) $$\begin{cases} \dfrac{\partial u_1}{\partial t} = \Delta u_1 - a_{11}u_1 \ , & in \quad \Omega \times [0,T] \\ \dfrac{\partial u_2}{\partial t} = -a_{22}u_2 + g(u_1) \ , & in \quad \Omega \times [0,T] \\ Bu_1 = H_w[u_2] \ , & in \quad \partial\Omega \times [0,T] \\ u_1(x,0) = u_1^o(x) \ , & x \in \Omega \\ u_2(x,0) = u_2^o(x) \ , & x \in \Omega \end{cases}$$

This problem has been faced by Arnautu et al. [13] for suitable choices of the cost functions f, h, l. An existence theorem for the optimal solution (u_1^*, u_2^*, w^*) of Problem (P) has been proven, together with the necessary conditions that it must satisfy. Numerical algorithms are provided for the concrete evaluation of the optimal solution.

For the simplified case

$$
\begin{aligned}
&f \equiv 0 \ , \quad l(u_2) = u_2 \ , \quad h(w) = \frac{1}{w^2} \\
&\Omega = [0,1] \times [0,1] \\
&\Gamma_1 = [0,1] \times \{0\} \\
&T = 1 \\
&g(u_1) = \beta \frac{u_1}{\gamma u_1 + 1} \\
&\beta = 50 \\
&\gamma = 2 \\
&u_1^o \neq 0 \qquad \text{in} \quad \Omega_S := [0.3, 0.7] \times [0.3, 0.7] \\
&u_1^o = 0 \qquad \text{in} \quad \Omega - \Omega_S \\
&u_2^o = 0 \qquad \text{in} \quad \Omega
\end{aligned}
$$

the optimal sanitation program is illustrated in Fig.7.1.

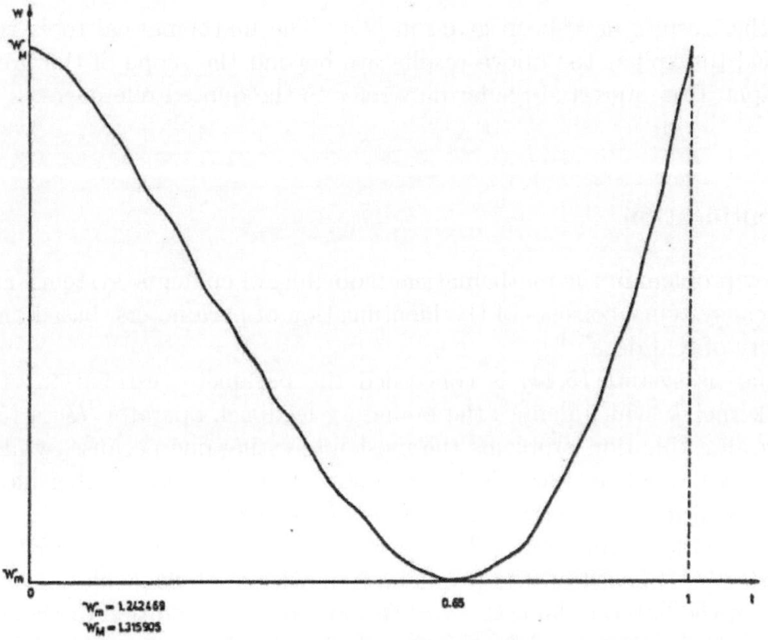

Fig. 7.1. [13]

An interesting modification of Problem (P) has been studied by Basile and Mininni [24]. The total duration T of the sanitation program is itself a parameter to be minimized under suitable constraints which impose upper bounds on the specific costs J_1 and J_2.

Problem (P'). *Minimize the cost*

$$J(T, u_1, u_2, w) := T \tag{7.7}$$

on all (u_1, u_2, w) *subject to the state system (7.6), under the constraints*

$$\begin{aligned} &\int_\Omega u_2(x,T)\,dx \leq \overline{u}_2 \\ &\int_0^T \int_{\partial\Omega} h(w(x,t))\,dx\,dt \leq c_1 \\ &\int_0^T \int_\Omega f(u_2(x,t))\,dx\,dt \leq c_2 \end{aligned} \tag{7.8}$$

Further results have been given in [25]. The mathematical tools which are needed to explicit the above results are beyond the scope of the present monograph. The interested reader may refer to the quoted references.

7.2. Identification

A key problem in the mathematical modelling of epidemic systems, as for all physical systems, consists of the identification of parameters, based on the availability of real data.

As far as system (5.44) is concerned the parameter estimation of the integral kernel k which defines the boundary feedback operator H of (5.46) is among all estimation problems the most interesting one because, while we do not have an explicit dynamical interpretation for it, on the other hand it plays a central role in the transmission of the disease.

A method for estimating k has been proposed and analyzed by Kunisch and Schelch [141] based on the (available) knowledge of the evolution of the epidemic in the habitat $u_2(x;t)$, $x \in \Omega$, and on the evolution of the agent along the sea shore, $u_1(x;t)$, $x \in \partial\Omega$, during an observation time interval $[0,T]$.

The analysis of the problems raised in this chapter require mathematical methods that go beyond the common background of an applied mathematician. So we suggest the interested reader to refer to the quoted literature.

Appendix A. Ordinary differential equations

A.1. The initial value problem for systems of ODE 's

We shall consider here dynamical systems defined by systems of ordinary differential equations of the form

$$\frac{dz}{dt} = f(t, z) \ , \tag{A.1}$$

with $f \in C(J \times D, \mathbb{R}^n)$, a continuous function of the variables $(t, z) \in J \times D$, where $J \subset \mathbb{R}$ is an open interval, and $D \subset \mathbb{R}^n$ is an open subset.

A function $z : J_z \to D$ is called a solution of the differential equation (A.1) (in J_z) if the following holds:

(i) $J_z \subset J$ is a nonempty interval

(ii) $z \in C^1(J_z, D)$, is continuous in J_z up to its first derivative

(iii) for any $t \in J_z$, $\dfrac{dz}{dt}(t) = f(t, z(t))$.

Remark. A function $z : J_z \to D$ is a solution of the differential equation (A.1) in J_z iff $z \in C(J_z, D)$ and for any $t_0 \in J_z$:

$$z(t) = z(t_0) + \int_{t_0}^{t} f(s, z(s))\, ds \ , \qquad t \in J_z \ . \tag{A.2}$$

We shall say that the function $f : J \times D \to \mathbb{R}^n$ satisfies the "Lipschitz condition" with respect to z if a constant $L > 0$ exists such that

$$\|f(t, z_1) - f(t, z_2)\| \leq L \, \|z_1 - z_2\|$$

for $t \in J$ and $z_1, z_2 \in D$. L is called the "Lipschitz constant".

Given $t_0 \in J$ and $z_0 \in D$ we say that f satisfies a "local Lipschitz condition" in (t_0, z_0) with respect to z if a, d, $L > 0$ constants exist such that

$$\|f(t, z_1) - f(t, z_2)\| \leq L \, \|z_1 - z_2\|$$

for $t \in [t_0 - a, t_0 + a]$, and $z_1, z_2 \in \overline{B}_d(z_0) := \{z \in \mathbb{R}^n \mid \|z - z_0\| \leq d\}$ (clearly a, d, L all depend upon (t_0, z_0)).

Remark. Note that f will always satisfy a local Lipschitz condition at any point of a domain $J \times D \subset \mathbb{R} \times \mathbb{R}^n$ whenever f and its partial derivatives $\partial f / \partial z_i$, $i = 1, \cdots, n$ are continuous in $J \times D$.

Theorem A.1. [85, 216] *Let* $f \in C(J \times D, \mathbb{R}^n)$, *be locally Lipschitzian in* $(t_0, z_0) \in J \times D$. *Then a* $\Delta > 0$ *exists such that the differential equation (A.1) admits a unique solution* $z \in C^1((t_0 - \Delta, t_0 + \Delta), D)$ *satisfying*

$$(A.1o) \qquad z(t_0) = z_0 \quad .$$

Remark. It can be further shown that, with the above notations,

$$\Delta = \min\left\{a, \frac{d}{M}\right\}$$

where

$$M := \sup_{\substack{|t-t_0| \le a \\ \|z-z_0\| \le d}} |f(t,z)|$$

The possibility of (global) existence of the solution of problem (A.1), (A.1*o*) in the whole "time" interval $J \subset \mathbb{R}$ is left to the following theorem.

Theorem A.2. [3] *Let* $f \in C(J \times D, \mathbb{R}^n)$ *be locally Lipschitzian in* $J \times D$. *Then for any* $(t_0, z_0) \in J \times D$ *there exists a unique nonextendible solution*

$$z(\cdot\,; t_0, z_0) \; : \; J(t_0, z_0) \longrightarrow D$$

of the initial value problem (A.1), (A.1o). The maximal interval of existence $J(t_0, z_0)$ is open:

$$J(t_0, z_0) := (\tau^-(t_0, z_0), \tau^+(t_0, z_0))$$

and we either have

$$\tau^- := \tau^-(t_0, z_0) = \inf J \;, \qquad \text{resp.} \qquad \tau^+ := \tau^+(t_0, z_0) = \sup J \;,$$

or

$$\lim_{t \to \tau^{\pm}} \min\left\{dist(z(t\,; t_0, z_0), \partial D) \;,\; |z(t\,; t_0, z_0)|^{-1}\right\} = 0 \;.$$

[Here of course we mean the limit as $t \to \tau^-$ when $\tau^- > \inf J$, and $t \to \tau^+$ when $\tau^+ < \sup J$, respectively. Moreover we use the convention $dist(x, \emptyset) = \infty$].

The above result can be expressed somewhat imprecisely as : either the solution exists for all time, or it approaches the boundary of D (where the boundary of D includes the "point at infinity" ($\|z\| = \infty$)).

A slight refinement of the above result is given by the following corollary.

Corollary A.3. [223] *Under the same assumption of Theorem A.2, assume further that given* $(t_0, z_0) \in J \times \mathbb{R}^n$ *there is a function* $m \in C([t_0, \sup J), \mathbb{R}_+)$ *such that, for any* $t \in [t_0, \tau^+(t_0, z_0))$:

$$\|z(t\,; t_0, z_0)\| \le m(t) \;,$$

then

$$\tau^+(t_0, z_0) = \sup J \ .$$

A useful criterion, implying the boundedness of all solutions of the differential equation (A.1) (for finite time), is given by the following proposition.

Proposition A.4. [3] *Assume there exist* $\alpha, \beta \in C(J, \mathbb{R}_+) \cap L^1(J, \mathbb{R})$ *such that*

$$\|f(t,z)\| \leq \alpha(t)\,\|z\| + \beta(t) \ , \qquad (t,z) \in J \times D \ .$$

Then for any $(t_0, z_0) \in J \times D$, we have $\tau^+(t_0, z_0) = \sup J$.

A particular case of Proposition A.4 is given by linear systems of differential equations.

Theorem A.5. [3, 223] *Let* $A \in C\left(J, \mathbb{R}^{n\times n}\right)$, *and* $b \in C(J, \mathbb{R}^n)$. *Then the linear (nonhomogeneous) IVP*

$$\dot{z} = A(t)\,z + b(t) \quad ; \quad z(t_0) = z_0 \tag{A.3}$$

has a unique global solution for every $(t_0, z_0) \in J \times \mathbb{R}^n$.

For any choice of $(t_0, z_0) \in J \times D$, Theorem A.2 defines a solution $\{z(t;\, t_0, z_0)\ ,\ t \in (\tau^-(t_0, z_0)\ ,\ \tau^+(t_0, z_0))\}$.

Now $z = z(t;\, t_0, z_0)$ may be seen as a function of the $2+n$ variables (t, t_0, z_0) ; its domain of definition is

$$\Omega := \bigcup_{(t_0,z_0)\in J\times D} I(t_0, z_0) \subset \mathbb{R}\times\mathbb{R}\times\mathbb{R}^n$$

where

$$I(t_0,z_0) := \left\{(t, t_0, z_0) \mid \tau^-(t_0,z_0) < t < \tau^+(t_0,z_0)\right\} \subset \mathbb{R}\times\mathbb{R}\times\mathbb{R}^n \ .$$

Theorem A.6. [223] *The domain of definition* Ω *of the (maximal) solution function* $z(t;\, t_0, z_0)$ *of system (A.1), (A.1o) is an open set in* $\mathbb{R}\times\mathbb{R}\times\mathbb{R}^n$ *and* z *is continuous on* Ω.

Further regularity of the solution function $z(t;\, t_0, z_0)$ is induced by the regularity of f [3, 223].

For any $(t_0, z_0) \in J \times D$, the set of points in $\mathbb{R}^{n+1}$ given by

$$\{(t\,,\, z(t;\, t_0, z_0)) \mid t \in I(t_0, z_0)\}$$

will be called the "trajectory" through (t_0, z_0).

The "path" or "orbit" of a trajectory is the projection of the trajectory into $\mathbb{R}^n$, the space of dependent variables in (A.1). The space of dependent variables is usually called the "state space" or "phase space" [104].

Suppose, for simplicity, that system (A.1), (A.1o) possesses a forward-unique solution $z(\cdot) = z(\cdot\,; t_0, z_0)$ defined on the forward interval $J(t_0) = [t_0, +\infty)$, for every $(t_0, z_0) \in \mathbb{R} \times D$. Then, defining $\varphi(\tau, t_0, z_0) := z(\tau + t_0, t_0, z_0)$, $\tau \in \mathbb{R}_+$, we see that the mapping $\varphi : \mathbb{R}_+ \times \mathbb{R} \times D \longrightarrow \mathbb{R}^n$ satisfies

(i) $\varphi(0\,; t_0, z_0) = z_0$, for any $(t_0, z_0) \in \mathbb{R} \times D$

(ii) $\varphi(\tau + s\,; t_0, z_0) = \varphi(\tau\,; s + t_0, \varphi(s\,; t_0, z_0))$, for any $\tau, s \in \mathbb{R}_+$, $(t_0, z_0) \in \mathbb{R} \times D$

(iii) φ is continuous on $\mathbb{R}_+ \times \mathbb{R} \times D$.

Because of (i), (ii), (iii), we say that system (A.1) generates a "process" φ [215, 223].

A.1.1. Autonomous systems

In this section we refer to the case in which the function f of system (A.1) does not depend explicitly upon t; namely we consider the "autonomous" system

$$(A.4) \qquad \frac{dz}{dt} = f(z)\ .$$

Now $f \in C(D)$, a continuous function of $z \in D$, with D an open subset of $\mathbb{R}^n$.

A basic property of autonomous systems is the following : if $z(t)$ is a solution of (A.4) on an interval $(a, b) \subset \mathbb{R}$, then, for any real number τ, the function $z(t - \tau)$ is a solution of (A.4) on the interval $(a + \tau, b + \tau)$. This is clear since the differential equation remains unchanged by a translation of the independent variable.

In particular we have

$$z(t\,; t_0, z_0) = z(t - t_0\,; 0, z_0)$$

for any $t \in J(t_0, z_0)$.

Moreover

$$J(t_0, z_0) = J(0, z_0) + t_0$$

for any $(t_0, z_0) \in J \times D$ (this means that $J(t_0, z_0)$ is obtained by shifting the interval $J(0, z_0)$ of the quantity t_0).

Thus the family of solutions of (A.4) subject to the initial condition

$$(A.4o) \qquad z(0) = z_0$$

completely defines the set of solutions of (A.4) subject to initial conditions of the more general form (A.1*o*).

For $z_0 \in D$ we now set

$$\begin{aligned} \tau^{\pm}(z_0) &:= \tau^{\pm}(0, z_0) \\ J(z_0) &:= J(0, z_0) = (\tau^{-}(z_0), \tau^{+}(z_0)) \end{aligned}$$

and we define

$$\varphi(t, z_0) := z(t\,;\, 0, z_0)$$

We now let

$$\Omega := \{(t, z) \in \mathbb{R} \times D \mid \tau^{-}(z) < t < \tau^{+}(z)\} \ .$$

By specializing Theorem A.6 to the autonomous case, Ω is an open subset of $\mathbb{R}^{n+1}$, and φ is continuous on Ω.

Moreover

(i) $\varphi(0; z_0) = z_0$, for any $z_0 \in D$

(ii) $\varphi(t+s\,;\, z_0) = \varphi(t\,;\, \varphi(s, z_0))$, for any $z_0 \in D$ and for all $s \in J(z_0)$, and $t \in J(\varphi(s\,;\, z_0))$.

Because of properties (i) and (ii) we say that system (A.4) generates a "(local) flow" or a "(local) dynamical system" on $D \subset \mathbb{R}^n$.

If $\Omega = \mathbb{R} \times D$, that is $\tau^{-}(z_0) = -\infty$ and $\tau^{+}(z_0) = +\infty$ for all $z_0 \in D$, then φ is called a "global flow" or a "global dynamical system".

For any $z_0 \in D$ we call

$$\begin{aligned} \Gamma_{+}(z_0) &:= \{\varphi(t\,;\, z_0) \mid t \in [0, \tau^{+}(z_0))\} \ , \\ \Gamma_{-}(z_0) &:= \{\varphi(t\,;\, z_0) \mid t \in (\tau^{-}(z_0), 0]\} \ , \\ \Gamma(z_0) &:= \{\varphi(t\,;\, z_0) \mid t \in J(z_0)\} \ , \end{aligned}$$

the "positive semiorbit", the "negative semiorbit", the "orbit" through z_0 respectively.

Note that each point of D belongs exactly to one orbit.

In the context of autonomous systems the family of vectors $\{f(z) \ , \ z \in D\}$ or better the family $\{(z, f(z)) \ , \ z \in D\}$ is called the "vector field" defining system (A.4).

A point $z^* \in D$ is called an "equilibrium point" or "critical point" or "singularity point" or "stationary point" of an n-dimensional vector field $f \in C^1(D, \mathbb{R}^n)$ if

$$f(z^*) = 0 \ .$$

If $z^* \in D$ is an equilibrium point then clearly

$$z(t) = z^* \ , \qquad t \in \mathbb{R}$$

is a (global) solution of system (A.4).

The trajectory of the critical point z^* is the line in $\mathbb{R}^{n+1}$ given by $\mathbb{R} \times \{z^*\}$. The orbit of a critical point z^* is $\Gamma(z^*) = \{z^*\} \subset \mathbb{R}^n$.

With the notations introduced above, the following proposition holds true.

Proposition A.7. [223] *Let $z_0 \in D$; if $\varphi(t\,;\,z_0)$ approaches a point a in D as $t \to \tau^+(z_0)$ then $\tau^+(z_0) = +\infty$ and a is a critical point.*

Let C_ψ denote a curve in $\mathbb{R}^n$ parametrized by a continuous function $\psi \in C([a,b], \mathbb{R}^n)$. If C_ψ does not intersect itself, that is, if ψ is one-to-one (injective) then it is called a "simple curve" or "arc".

If $\psi(a) = \psi(b)$, but still $\psi(t_1) \neq \psi(t_2)$ for every $t_1 \neq t_2$ in $[a, b)$, then C_ψ is called a "simple closed curve" or "Jordan curve".

Theorem A.8. [104, 223] *Let C denote the orbit of the autonomous system (A.4). The following statements are equivalent*

(i) C intersects itself in at least one point.
(ii) C is the orbit of a periodic solution.
(iii) C is a simple, closed (Jordan) curve.

Because of Theorem A.8 an orbit which is a Jordan curve is sometimes called a "periodic orbit".

Note that if Γ is a closed orbit of (A.4) and $z_0 \in \Gamma$, there is a $\tau \neq 0$ such that $\varphi(\tau\,;\,z_0) = z_0 = \varphi(0\,;\,z_0)$. By uniqueness of solutions $\varphi(t+\tau\,;\,z_0) = \varphi(t\,;\,z_0)$ for all $t \in \mathbb{R}$, which says that $\{\varphi(t\,;\,z_0)\ ,\ t \in \mathbb{R}\}$ has period τ.

If Γ is a closed path, not reducing to an equilibrium point, there exists a smallest positive period $T > 0$, which is called the "minimal" or "fundamental period" of $\varphi(t\,;\,z_0)$ for any $z_0 \in \Gamma$; i.e., any $z_0 \in \Gamma$ is a fixed point of the map $\varphi(T\,;\,\cdot) : D \longrightarrow D$.

The map $\varphi(T\,;\,\cdot)$ is usually called the "monodromy operator" of our dynamical system.

It can be shown [3] that there exist exactly three types of orbits for the autonomous system (A.4) in $\mathbb{R}^n$: (i) stationary points; (ii) periodic orbits; (iii) "open" simple curves. Clearly in cases (i) and (ii) $\tau^+(z_0) = +\infty$ and $\tau^-(z_0) = -\infty$, for any point z_0 of the orbit.

A.1.1.1. Autonomous systems. Limit sets, invariant sets

In this section we consider system (A.4) and assume that f satisfies enough conditions on D, an open subset of the space $\mathbb{R}^n$, to ensure that there exists a unique global solution $\{\varphi(t\,;\,z_0)\;,\;t \in \mathbb{R}\}$ for any $z_0 \in D$.

A point $p \in \mathbb{R}^n$ is called an "ω-limit point" of the solution $\varphi(t\,;\,z_0)$ iff there is a sequence $(t_k)_{k\in\mathbb{N}}$ of times such that

$$(i) \quad \lim_{k\to\infty} t_k = +\infty$$

$$(ii) \quad \lim_{k\to\infty} \varphi(t_k\,;\,z_0) = p$$

Similarly a point $q \in \mathbb{R}^n$ is called an "α-limit point" of the solution $\varphi(t\,;\,z_0)$ iff there is a sequence of real numbers $(t_k)_{k\in\mathbb{N}}$ such that

$$(i)' \quad \lim_{k\to\infty} t_k = -\infty$$

$$(ii)' \quad \lim_{k\to\infty} \varphi(t_k\,;\,z_0) = q\;.$$

The "ω-limit set" of an orbit Γ associated with the dynamical system φ is the set of all its ω-limit points

$$\omega(\Gamma) := \bigcap_{z_0\in\Gamma} \overline{\Gamma_+(z_0)}\;.$$

Similarly the "α-limit set" of an orbit Γ associated with the dynamical system φ is the set of all its α-limit points

$$\alpha(\Gamma) := \bigcap_{z_0\in\Gamma} \overline{\Gamma_-(z_0)}$$

A set $M \subset \mathbb{R}^n$ is called an "invariant set" of (A.4) if, for any $z_0 \in M$, the solution $\{\varphi(t\,;\,z_0)\;,\;t \in \mathbb{R}\}$ belongs to M :

$$\varphi(t\,;\,M) \subset M\;, \quad \text{for any} \quad t \in \mathbb{R}\;.$$

Any orbit is obviously an invariant set of (A.4).

A set $M \subset \mathbb{R}^n$ is called "positively (negatively) invariant" if for each $z_0 \in M$, $\{\varphi(t\,;\,z_0)\;,\;t \in \mathbb{R}_+\}$ ($\{\varphi(t\,;\,z_0)\;,\;t \in \mathbb{R}_-\}$) belongs to M.

Theorem A.9. [104] *The α- and ω-limit sets of an orbit Γ are closed and invariant. Furthermore if for some $z_0 \in D$, $\Gamma_+(z_0)$ (resp. $\Gamma_-(z_0)$) is bounded, then $\omega(\Gamma(z_0))$ (resp. $\alpha(\Gamma(z_0))$) is nonempty, compact and connected. Moreover $dist(\varphi(t\,;\,z_0)\;,\;\omega(\Gamma(z_0))) \longrightarrow 0\;,\quad as\ t \to +\infty$ (resp. $dist(\varphi(t\,;\,z_0)$, $\alpha(\Gamma(z_0))) \longrightarrow 0\;,\;as\ t \to -\infty$).*

A.1.1.2. Two-dimensional autonomous systems

We specialize to the case of a two-dimensional autonomous system

$$(A.5) \qquad \begin{cases} \dfrac{dz_1}{dt} = f_1(z_1, z_2) \\ \dfrac{dz_2}{dt} = f_2(z_1, z_2) \ , \end{cases}$$

where $f := (f_1, f_2)$ is a real continuous vector function defined on a bounded open subset D of the real plane $\mathbb{R}^2$.

We assume further that for each real $t_0 \in \mathbb{R}$ and each point $z_0 \in D$ there exists a unique solution $\varphi(t\,;\, z_0)$ of (A.5) such that $\varphi(t_0\,;\, z_0) = z_0$.

The theory ensures that φ is a continuous function of (t_0, z_0) for all t for which φ is defined, and for all $z_0 \in D$.

For autonomous systems with phase space $D \subset \mathbb{R}^2$, it can be proved that every limit set is a critical point, a closed path, or a combination of solution paths and critical points joined together. In case $n = 2$ the phase space is often called the "Poincaré phase plane".

Theorem A.10. [223] *If the ω-limit set Ω for a solution of the system (A.5) contains a closed path C, then $\Omega = C$.*

Theorem A.11. [223] *If a path C of system (A.5) contains one of its own ω-limit points, then C is either a critical point or a closed path.*

Proposition A.12. *If Γ_+ and $\omega(\Gamma_+)$ have a regular point in common, then Γ_+ is a periodic orbit.*

It is clear that if the ω-limit set of a solution consists of precisely one closed path C, then the ω-limit set contains no critical point.

The converse of this statement is contained in the following

Theorem A.13. (Poincaré-Bendixson) [104] *If Γ_+ is a bounded positive semiorbit and $\omega(\Gamma_+)$ does not contain a critical point, then either*

(i) $\Gamma_+ = \omega(\Gamma_+)$

or

(ii) $\omega(\Gamma_+) = \overline{\Gamma_+} - \Gamma_+$

In either case the ω-limit set is a periodic orbit, and in the latter case it is referred to as a limit cycle.

Hence, a closed path which is a limit set of a path other than itself is called a "limit cycle".

If a closed path is a limit cycle, then an other path must approach it spirally.

Theorem A.14. [104] *A closed path of system (A.5) must have a critical point in its interior.*

Another important test for the non existence of periodic solutions is provided by the following

Theorem A.15. (Bendixson negative criterion) [210] *Let f in system (A.5) have continuous first partial derivatives on an open simply connected subset $D \subset \mathbb{R}^2$. System (A.5) cannot have periodic solutions in D if $div f = \frac{\partial}{\partial x} f_1 + \frac{\partial}{\partial y} f_2$ has the same sign throughout D.*

The Bendixson negative criterion can be extended in the following form.

Theorem A.16. (Bendixson-Dulac criterion) [216] *Under the same assumptions as in Theorem A.15, suppose further that there exists a continuously differentiable function $\beta : D \longrightarrow \mathbb{R}$ such that the function $\frac{\partial}{\partial x}(\beta f_1) + \frac{\partial}{\partial y}(\beta f_2)$ does not change sign in D. Then there are no periodic solutions of system (A.5) in the region D.*

A.2. Linear systems of ODE 's

A.2.1. General linear systems

Consider the linear system of $n \in \mathbb{N} - \{0\}$ first order equations

$$(A.6) \qquad \frac{d}{dt} z_j = \sum_{k=1}^{n} a_{jk}(t)\, z_k + b_j(t) \ , \qquad j = 1, \cdots, n$$

where the a_{jk} and b_j for $j, k = 1, \cdots, n$ are continuous real valued functions on an interval $J \subset \mathbb{R}$. In matrix notation, system (A.6) can be rewritten as

$$(A.7) \qquad \frac{dz}{dt} = A(t)\, z + b(t)$$

where $A(t) = (a_{jk}(t))$, $j, k = 1, \cdots, n$ and $b(t) = (b_1(t), \cdots, b_n(t))^T$.

Thanks to Theorem A.5, system (A.7) admits a unique global solution in J, subject to the initial condition

$$(A.7o) \qquad z(t_0) = z_0$$

for any choice of $(t_0, z_0) \in J \times \mathbb{R}^n$.

The basic characteristic property of linear systems of the form (A.7) is the so called "Principle of Superposition": If $z(t)$ is any solution of (A.7) corresponding to the "forcing term" $b(t)$, and $y(t)$ is a solution of (A.7) corresponding to the forcing term $h(t)$, then for any choice of real (or complex) numbers c and d, $c\, z(t) + d\, y(t)$ is a solution of (A.7) corresponding to the forcing term $c\, b(t) + d\, h(t)$.

In particular if $b = h$, $z(t) - y(t)$ is a solution of the homogeneous system

$$(A.8) \qquad \frac{dz}{dt} = A(t)\, z \ .$$

Thus if $z(t)$ is a solution of (A.8) and $z_p(t)$ is a solution of (A.7) then $z(t) + z_p(t)$ is again a solution of (A.7).

We may then confine our analysis in the search of the "general" solution of the homogeneous system (A.8).

Theorem A.17. [104, 216] *Every linear combination of solutions of (A.8) is a solution of (A.8).*

Suppose now that we have n solution vectors $z^{(i)}(t)$, $i = 1, \cdots, n$ of (A.8) defined on $J \subset \mathbb{R}$.

We can form a matrix $X(t)$ whose columns are these solutions :

$$X(t) := \left[z^{(1)}(t), \cdots, z^{(n)}(t) \right] \ , \qquad t \in J \ .$$

Clearly, in J,

$$\dot{X}(t) = A(t)\,X(t)\ . \tag{A.9}$$

Furthermore, if $c \in \mathbb{R}^n$ is any constant vector, then $X(t)\,c$ is a solution of (A.8).

Lemma A.18. [104] *If $X(t)$ is a $n \times n$ matrix solution of (A.9), then either $detX(t) = 0$, for all $t \in J$, or $detX(t) \neq 0$ for all $t \in J$.*

An $n \times n$ matrix $X(t)$ solution of (A.9) such that $detX(t) \neq 0$ for all $t \in J$, will be called a "fundamental matrix solution" of system (A.7).

Theorem A.19. [104, 216] *If $X(t)$ is any fundamental matrix solution of (A.8) in $J \subset \mathbb{R}$, then every solution of (A.8) in J can be written as $X(t)\,c$ for an appropriate constant vector $c \in \mathbb{R}^n$.*

We shall say that $\{z(t;c) = X(t)\,c\ ;\ c \in \mathbb{R}^n\}$ is a "general integral" of system (A.8).

In order to obtain the solution passing through the point $(t_0, z_0) \in J \times \mathbb{R}^n$, we need to choose $c = X^{-1}(t_0)\,z_0$, so that we get

$$z(t\,;\,t_0, z_0) = X(t)\,X^{-1}(t_0)\,z_0\ , \qquad t \in J \tag{A.10}$$

Remark. In order to obtain a fundamental matrix solution of system (A.8) it suffices to select n linearly independent initial vectors $z_0{}^{(i)}$, $i = 1, \cdots, n$, and finding the corresponding n linearly independent solutions of system (A.8).

If we now go back to the nonhomogeneous system (A.7) we may state that if $z_p(t)$ is a particular solution of (A.7), the general solution of (A.7) is given by

$$\{z(t;c) = X(t)\,c + z_p(t)\ ,\quad c \in \mathbb{R}^n\}$$

provided $X(t)$ is a fundamental matrix solution of (A.8).

Theorem A.20. [104] *If $X(t)$ is a fundamental matrix solution of (A.8) then every solution of (A.7) is given by the formula*

$$z(t) = X(t)\left[X^{-1}(\tau)\,z(\tau) + \int_\tau^t X^{-1}(s)\,b(s)\,ds\right]$$

for any choice of $\tau \in J$.

The above theorem (known as the "variation of constants formula") gives us in particular the solution passing through the point $(t_0, z_0) \in J \times \mathbb{R}^n$:

$$z(t\,;\,t_0, z_0) = X(t)\,X^{-1}(t_0)\,z_0 + \int_{t_0}^t X(t)\,X^{-1}(s)\,b(s)\,ds \tag{A.11}$$

Note that $X(t,t_0) := X(t)\,X^{-1}(t_0)$ is such that $X(t_0,t_0) = I$. It is the unique solution of (A.9) such that this happens. The matrix $X(t,t_0)$ is known as the "principal matrix solution" of (A.7) at initial time $t_0 \in J$.

It can be easily shown that

$$X(t,\tau) = X(t,s)\,X(s,\tau) \tag{A.12}$$

for any choice of τ, s, t in J. Thus the variation of constants formula (A.11) can be rewritten as

$$z(t\,;\,t_0,z_0) = X(t,t_0)\,z_0 + \int_{t_0}^{t} X(t,s)\,b(s)\,ds. \tag{A.13}$$

A.2.2. Linear systems with constant coefficients

In this section we consider the homogeneous equation

$$\dot{z} = Az \tag{A.14}$$

where A is an $n \times n$ real constant matrix. In this case we may assume $J = \mathbb{R}$.

Since system (A.14) is autonomous we may reduce (A.11) to considering only the case $t_0 = 0$:

$$z(t;z_0) = X(t)\,X^{-1}(0)\,z_0 + \int_0^t X(t)\,X^{-1}(s)\,b(s)\,ds\ ,\ t \in \mathbb{R}\ . \tag{A.15}$$

The principal matrix solution of (A.14) $P(t) := X(t,0)$ is such that $P(0) = I$. In this case equation (A.12) becomes

$$P(t+s) = P(t)\,P(s) \tag{A.16}$$

for any s, $t \in \mathbb{R}$.

This relation suggests the notation e^{At}, for $P(t)$, $t \in \mathbb{R}$.

As a consequence of the definition the $n \times n$ matrix e^{At} , $t \in \mathbb{R}$ satisfies the following properties

(i) $e^{A0} = I$

(ii) $e^{A(t+s)} = e^{At}\,e^{As}\ ,\quad s,\, t \in \mathbb{R}$

(iii) $\left(e^{At}\right)^{-1} = e^{-At}\ ,\quad t \in \mathbb{R}$

(iv) $\dfrac{d}{dt}e^{At} = A\,e^{At} = e^{At}\,A\ ,\quad t \in \mathbb{R}$

(v) a general solution of (A.14) is $\{e^{At}c\ ,\ c \in \mathbb{R}^n\}$

(vi) if $X(t)$ $(detX(0) \neq 0)$ is a fundamental matrix solution of (A.14) then

$$e^{At} = X(t)\,X^{-1}(0)\ .$$

With these notations, equation (A.15) may be rewritten as

$$(A.17) \qquad z(t;z_0) = e^{At}\,z_0 + \int_0^t e^{A(t-s)}\,b(s)\,ds \qquad t \in \mathbb{R}$$

In addition to the above properties e^{At} can be obtained as the sum of a convergent power series of matrices which resembles structurally the scalar case [216]

$$(A.18) \qquad e^{At} = \sum_{k=0}^{\infty} \frac{(A\,t)^k}{k\,!}$$

Note that for any $t \in \mathbb{R}$, e^{At} is always nonsingular, i.e. $det\left(e^{At}\right) \neq 0$.

But, what is an effective means for computing e^{At} ? To find it we need to introduce the concepts of eigenvalue and eigenvector of a matrix [38, 3].

A complex number λ is called an "eigenvalue" of an $n \times n$ real matrix A if there exists a nonzero vector v such that

$$(A.19) \qquad (A - \lambda I)\,v = 0$$

Any nonzero solution $v \in \mathbb{R}^n$ of equation (A.19) is called an "eigenvector" associated with the eigenvalue λ.

It is well known that (A.19) admits (for a fixed λ) a nontrivial solution v iff the matrix $(A - \lambda I)$ is singular. Thus λ needs to satisfy the "characteristic equation"

$$(A.20) \qquad det(A - \lambda I) = 0\ .$$

This equation is a (real) polynomial of degree n in λ and, therefore, admits n solutions in **C**, not all of which may be distinct.

On the other hand, if $\lambda_1, \cdots, \lambda_k$ are $k \leq n$ distinct eigenvalues of the matrix A and $v^1, \cdots, v^k$ are corresponding eigenvectors, then $v^1, \cdots, v^k$ are linearly independent.

The set of all eigenvalues of A is called the "spectrum" of A and is denoted by $\sigma(A)$.

Lemma A.21. *If λ is an eigenvalue of the matrix A and v is an eigenvector associated with λ, then the function*

$$z(t) = e^{\lambda t}\,v\ , \qquad t \in \mathbb{R}$$

is a solution of system (A.14).

If now A admits n distinct eigenvalues $\lambda_1, \cdots, \lambda_n$ and $v^1, \cdots, v^n$ are corresponding eigenvectors, we may claim, as stated above, that these eigenvectors are linearly independent.

We may then proceed as suggested in the Remark following Theorem A.19.

Let $V(t)$ be the solution of (A.9) (with $A(t) \equiv A$) such that $V(0) = [v^1, \cdots, v^n]$. Due to Lemma A.21, we have

$$V(t) = \left[e^{\lambda_1 t} v^1, \cdots, e^{\lambda_n t} v^n\right] ,$$

and since $V(0)$ is nonsingular, $V(t)$ is a fundamental matrix of (A.14).

Thus a general solution of (A.14) is given by

$$z(t; c) = V(t)\, c \ , \qquad c \in \mathbb{R}^n$$

and, due to (vi)

$$e^{At} = V(t)\, V^{-1}(0) \ .$$

If the eigenvalues are not all distinct, it may be still possible to determine a set of n linearly independent eigenvectors of A and proceed as before to find a fundamental matrix of (A.14).

However the technique will fail if the eigenvectors do not form a basis for $\mathbb{R}^n$. Then there may be solutions of (A.14) which cannot be expressed in terms of only exponential functions and constant vectors.

To determine the form of e^{At} when A is an $n \times n$ arbitrary real matrix, compute the distinct eigenvalues $\lambda_1, \cdots, \lambda_k$ $(k \leq n)$ of A, with respective multiplicities $n_1, \cdots, n_k$, such that $n_1 + \cdots + n_k = n$.

Corresponding to each eigenvalue λ_j of multiplicity n_j consider the system of linear equations

$$(A.21) \qquad (A - \lambda_j I)^{n_j}\, v = 0 \ , \qquad j = 1, \cdots, k \ .$$

The linear algebraic system (A.21) has n_j linearly independent solutions, that span a subspace X_j of $\mathbb{R}^n$ $(j = 1, \cdots, k)$.

Moreover, for every $z \in \mathbb{R}^n$ there exists a unique set of vectors $z_1, \cdots, z_k$, with $z_j \in X_j$, such that

$$(A.22) \qquad z = z_1 + z_2 + \cdots + z_k \ .$$

In the language of linear algebra, this means that $\mathbb{R}^n$ can be represented as the direct sum of subspaces X_j , $j = 1, \cdots, k$

$$\mathbb{R}^n = X_1 \oplus \cdots \oplus X_k \ .$$

Each of the subspaces X_j is invariant under A, that is $A\, X_j \subset X_j$, $j = 1, \cdots, n$, hence, it is invariant under e^{At} and for system (A.14).

We look for a solution of system (A.14) subject to the initial condition $z(0) = z_0 \in \mathbb{R}^n$.

We know that

$$z(t; z_0) = e^{At} z_0 ,$$

but our objective is to give an explicit representation of $e^{At} z_0$.

In accordance with (A.22) there will exist a unique set of vectors $v_1, \cdots, v_k$, $v_j \in X_j$, such that

$$z_0 = v_1 + \cdots + v_k .$$

Hence

$$e^{At} z_0 = \sum_{j=1}^{k} e^{At} v_j = \sum_{j=1}^{k} e^{\lambda_j t} e^{(A-\lambda_j I)t} v_j .$$

Now, by using the series expansion of e^{At}

$$e^{(A-\lambda_j I)t} v_j = \left[I + t(A - \lambda_j I) + \frac{t^2}{2!}(A - \lambda_j I)^2 + \cdots + \frac{t^{n_j - 1}}{(n_j - 1)!}(A - \lambda_j I)^{n_j - 1} \right] v_j$$

since, because of (A.21), all other terms will vanish.

Thus, for any $t \in \mathbb{R}$,

$$z(t; z_0) = e^{At} z_0$$

$$= \sum_{j=1}^{k} e^{\lambda_j t} \left[\sum_{i=0}^{n_j - 1} \frac{t^i}{i!} (A - \lambda_j I)^i \right] v_j \tag{A.23}$$

We note that it may happen that a $q_j < n_j$ exists such that

$$(A - \lambda_j I)^{q_j} = 0 \qquad (j = 1, \cdots, n) ;$$

in such case the corresponding sum in (A.23) will contain only q_j terms instead of n_j terms.

We are now in a position to state the following

Theorem A.22. [38] *Given the linear system (A.14) if $\rho \in \mathbb{R}$ is such that*

$$\rho > \max_{\lambda \in \sigma(A)} \mathcal{R}e\, \lambda$$

then there exists a constant $k > 0$ such that for any $z_0 \in \mathbb{R}^n$:

$$\| e^{At} z_0 \| \leq k\, e^{\rho t} \| z_0 \| , \qquad t \in \mathbb{R}_+ ,$$

or simply

$$\| e^{At} \| \leq k\, e^{\rho t} , \qquad t \in \mathbb{R}_+ .$$

Remark. In Theorem A.22, the constant ρ may be chosen as any number greater than or equal to the largest of $\mathcal{R}e\ \lambda$, $\lambda \in \sigma(A)$, whenever every eigenvalue whose real part is equal to this maximum is itself simple. In particular, this is always true if A has no multiple eigenvalues.

Corollary A.23. *If all eigenvalues of A have negative real parts, then there exist constants $k > 0$, $\sigma > 0$ such that, for any $z_0 \in \mathbb{R}^n$*

$$\|e^{At} z_0\| \leq k\, e^{-\sigma t} \|z_0\| \ , \qquad t \in \mathbb{R}_+$$

or simply

$$\|e^{At}\| \leq k\, e^{-\sigma t} \ , \qquad t \in \mathbb{R}_+ \ .$$

Corollary A.24. *If all eigenvalues of A have real part negative or zero and if those eigenvalues with zero real part are simple, then there exists a constant $k > 0$ such that*

$$\|e^{At}\| \leq k \ , \qquad t \in \mathbb{R}_+$$

i.e., for any $z_0 \in \mathbb{R}^n$,

$$\|e^{At} z_0\| \leq k \|z_0\| \ , \qquad t \in \mathbb{R}_+ \ ,$$

hence any solution is bounded on $\mathbb{R}_+$.

A.3. Stability

Suppose $f : \mathbb{R} \times D \longrightarrow \mathbb{R}^n$, $D \subset \mathbb{R}^n$ be such to ensure existence, uniqueness and continuous dependence on parameters; e.g. $f \in C^1(\mathbb{R} \times D, \mathbb{R}^n)$.

Consider the (nonautonomous) system of ODE's

$$\text{(A.24)} \qquad \frac{dz}{dt} = f(t, z) \ ,$$

and let $\phi(t)$ be some solution of (A.24) existing for $t \in \mathbb{R}$.

As usual we shall denote by $z(t\,;\,t_0, z_0)$, a solution of (A.24) passing through $z_0 \in D$ at time $t_0 \in \mathbb{R}$ ($z(t_0\,;\,t_0, z_0) = z_0$).

The solution $\{\phi(t)\ ,\ t \in \mathbb{R}\}$ of (A.24) is said to be "stable" if for every $\epsilon > 0$ and every $t_0 \in \mathbb{R}$ there exists a $\delta > 0$ (δ may depend upon ϵ and t_0) such that whenever $\|z_0 - \phi(t_0)\| < \delta$, the solution $z(t\,;\,t_0, z_0)$ exists for any $t > t_0$ and satisfies

$$\|z(t\,;\,t_0, z_0) - \phi(t)\| < \epsilon \ , \qquad \text{for} \quad t \geq t_0 \ .$$

This definition may be extended as follows.

The solution $\{\phi(t)\ ,\ t \in \mathbb{R}\}$ is said to be "asymptotically stable" if it is stable and if there exists a $\delta_0 > 0$ (δ_0 may again depend upon t_0) such that whenever $\|z_0 - \phi(t_0)\| < \delta_0$, the solution $z(t\,;\,t_0, z_0)$ approaches $\phi(t)$ as t tends to $+\infty$; i.e.

$$\lim_{t\to+\infty} \|z(t\,;\,t_0, z_0) - \phi(t)\| = 0 \ .$$

In the same context as above, we speak of "uniform stability" of ϕ if δ does not depend upon t_0; we speak of "uniform asymptotic stability" if also δ_0 does not depend upon t_0.

Suppose now that f does not depend upon t, so that we consider the autonomous system

$$\frac{dz}{dt} = f(z) \ . \qquad (A.25)$$

Let $z^* \in D$ be a critical point of f $(f(z^*) = 0)$, hence an equilibrium of (A.25). We may specialize the previous definitions to the stability of z^*.

The equilibrium solution z^* of (A.25) is said to be "stable" if for each number $\epsilon > 0$ we can find a number $\delta > 0$ (depending upon ϵ) such that whenever $\|z_0 - z^*\| < \delta$, the solution $z(t; z_0)$ exists for all $t \geq 0$ and $\|z(t; z_0) - z^*\| < \epsilon$ for $t \geq 0$.

The equilibrium solution z^* is said to be asymptotically stable if it is stable and if there exists a number $\delta_0 > 0$ such that whenever $\|z_0 - z^*\| < \delta_0$, then $\lim_{t\to+\infty} \|z(t; z_0) - z^*\| = 0$.

The equilibrium solution z^* is said to be unstable if it is not stable.

Note that because system (A.25) is autonomous the numbers δ , δ_0 are independent of the initial time which can always be chosen to be $t_0 = 0$.

Thus the stability and asymptotic stability are always uniform.

The above definitions raise the problem of finding the region of asymptotic stability of a solution $\phi(t)$, i.e. the subset $\tilde{D} \subset D$ such that, for a given t_0, for all $z_0 \in \tilde{D}$ one has

$$\lim_{t\to\infty} \|z(t\,;\,t_0, z_0) - \phi(t)\| = 0 \ .$$

We shall speak of "global asymptotic stability in $\tilde{D}$" when $\tilde{D}$ is a known subset of D. When $\tilde{D} = D$ we shall simply say "global asymptotic stability".

A.3.1. Linear systems with constant coefficients

Let A be a real $n \times n$ matrix and consider the system

$$\frac{dz}{dt} = Az \ . \tag{A.26}$$

Clearly the zero solution is an equilibrium solution of (A.26).

It is an immediate consequence of Theorem A.22 and Corollaries A.23, A.24, the following

Theorem A.25. [38, 216] *If all the eigenvalues of A have nonpositive real parts and all those eigenvalues with zero real parts are simple, then the solution $z^* = 0$ of (A.26) is stable. If and only if all eigenvalues of A have negative real parts, the zero solution of (A.26) is asymptotically stable. If one or more eigenvalues of A have a positive real part, the zero solution of (A.26) is unstable.*

The reader should observe that actually, for linear systems, the stability properties are global in $\mathbb{R}^n$.

A.3.2. Stability by linearization

Theorem A.26. (Poincaré-Lyapunov) [210] *Consider the equation*

$$\frac{dz}{dt} = Az + B(t)\, z + f(t,z) \ . \tag{A.27}$$

Let A be a real constant $n \times n$ matrix with all eigenvalues having negative real parts; $B(t)$ is a continuous real $n \times n$ matrix with the property

$$\lim_{t \to +\infty} \|B(t)\| = 0 \ ;$$

the vector function $f \in C(J \times D, \mathbb{R}^n)$ is Lipschitz continuous in a neighborhood of $0 \in D$, an open subset of $\mathbb{R}^n$; it is such that

$$\lim_{\|z\| \to 0} \frac{\|f(t,z)\|}{\|z\|} = 0 \ , \quad \text{uniformly in} \quad t \in \mathbb{R} \ . \tag{A.28}$$

Then the solution z^* of (A.27) is asymptotically stable.

Note that in this case we do not necessarily have global asymptotic stability.

Theorem A.27. [210] *Under the same assumptions of Theorem A.26 if now A admits at least one eigenvalue with positive real part, then the trivial solution of (A.27) is unstable.*

Theorems A.26 and A.27 play a central role in the analysis of the local behavior of a nonlinear autonomous system.

Suppose we are given the system

$$(A.29) \qquad \frac{dz}{dt} = F(z)$$

where $F \in C^1(D)$, D an open subset of $\mathbb{R}^n$.

Suppose that $z^* \in D$ is an equilibrium of (A.29) so that $F(z^*) = 0$.

If $\{z(t)\ ,\ t \in \mathbb{R}_+\}$ is any solution of (A.29) we may write it in the form

$$z(t) = z^* + y(t)\ , \qquad t \in \mathbb{R}_+$$

so that

$$\begin{aligned}\frac{d}{dt}y(t) &= F(z^* + y(t)) \\ &= F(z^*) + J_F(z^*)\, y(t) + g(y(t)) \\ &= J_F(z^*)\, y(t) + g(y(t))\end{aligned}$$

with $g(0) = 0$. Here $J_F(z^*)$ is the Jacobi matrix of F at z^*, i.e. the constant matrix whose (i,j) elements are

$$\frac{\partial F_i}{\partial z_j}(z^*)\ , \qquad i,j = 1, \cdots, n\ .$$

Thus the displacement $y(t)$ of $z(t)$ from the equilibrium z^* satisfies the equation

$$(A.30) \qquad \frac{dy}{dt} = J_F(z^*)\, y + g(y)$$

which is of the form (A.27).

A.4. Quasimonotone (cooperative) systems

In this section we shall deal with systems of ODE's of the form

$$(A.31) \qquad \frac{dz}{dt} = f(t, z)$$

where $f \in C^1\left(J \times \mathbb{R}^n_+\ ,\ \mathbb{R}^n\right)$ is a quasimonotone (cooperative) function for any $t \in J \subset \mathbb{R}$.

This means that, for any $t \in J$, the off-diagonal terms of the Jacobi matrix $J_f(t; z)$ at any point $z \in \mathbb{R}^n_+$ are nonnegative

$$\frac{\partial f_i}{\partial z_j}(t; z) \geq 0\ , \qquad i \neq j$$

for $i, j = 1, \cdots, n$.

A.4.1. Quasimonotone linear systems

Notations. Let $\mathbb{K}$ denote the positive cone of $\mathbb{R}^n$, i.e. its nonnegative orthant,

$$\mathbb{K} := \mathbb{R}^n_+ := \{z \in \mathbb{R}^n \mid z_i \geq 0 \ , \ i = 1, \cdots, n\} \ .$$

This cone induces a partial order in $\mathbb{R}^n$ via

$$y \leq x \qquad \text{iff} \qquad x - y \in \mathbb{K} \ .$$

In addition we shall use the notation

$$y < x \qquad \text{iff} \qquad x - y \in \mathbb{K} \ , \qquad \text{and} \quad x \neq y$$

$$y \ll x \qquad \text{iff} \qquad x - y \in \overset{\circ}{\mathbb{K}} \quad (\text{the interior of } \mathbb{K}) \ .$$

A nonnegative matrix B is a matrix with nonnegative entries $b_{ij} \geq 0 \ , \ i,j = 1, \cdots, n$. It is such that it leaves $\mathbb{K}$ invariant:

$$B(\mathbb{K}) \subset \mathbb{K}$$

A positive matrix B is a matrix with positive entries $b_{ij} > 0 \ , \ i,j = 1, \cdots, n$.

It satisfies the equivalent property that

$$B(\mathbb{K} - \{0\}) \subset \overset{\circ}{\mathbb{K}} \ .$$

In this section we shall analyze linear systems

$$\frac{dz}{dt} = Az \tag{A.32}$$

where A is a (nonnegative) "quasimonotone" $n \times n$ real matrix, i.e. such that all nondiagonal elements of A are nonnegative:

$$a_{ij} \geq 0 \ , \qquad i \neq j \ ; \ i,j = 1, \cdots, n \ . \tag{A.33}$$

Clearly for a certain $\alpha > 0$ the matrix $B = A + \alpha I$ will be a nonnegative matrix, i.e. all its elements will be nonnegative.

Theorem A.28. [93] *A necessary and sufficient condition that all the eigenvalues of the quasi-monotone matrix A should have negative real parts, is that the following inequalities be satisfied*

$$\begin{aligned} & a_{11} < 0 \\ & det \begin{pmatrix} a_{11} & a_{12} \\ a_{21} & a_{22} \end{pmatrix} > 0 \\ & \cdot \\ & \cdot \\ & \cdot \\ & (-1)^n \ det A > 0 \end{aligned} \tag{A.34}$$

Theorem A.29. [31, 93] *A quasimonotone matrix A always admits a real eigenvalue μ such that for any other $\lambda \in \sigma(A)$*

$$\mu \geq \mathcal{R}e\ \lambda\ ;$$

thus

$$\mu = \max\{\mathcal{R}e\ \lambda \mid \lambda \in \sigma(A)\}\ .$$

To this dominant eigenvalue μ there corresponds a nonnegative eigenvector $\eta \in \mathbb{K}$

$$A\eta = \mu\eta\ .$$

An $n \times n$ matrix B is "reducible" if for some permutation matrix P

$$P\,B\,P^T = \begin{pmatrix} B_1 & C \\ 0 & B_2 \end{pmatrix}\ ,$$

where B_1 and B_2 are square matrices. Otherwise B will be said "irreducible".

Theorem A.30. (Perron-Frobenius) [31, 93] *If A is an irreducible quasimonotone matrix its dominant eigenvalue μ is a simple eigenvalue of A. To μ there corresponds a positive eigenvector*

$$\eta \in \overset{\circ}{\mathbb{K}} \qquad (\eta \gg 0)\ .$$

Theorem A.31. [31] *Let A be a quasi-monotone matrix. Then*

(i) $$e^{At}\,\mathbb{K} \subset \mathbb{K}\ , \quad \text{for any} \quad t \geq 0\ .$$

Moreover

(ii) $$e^{At}\,(\mathbb{K} - \{0\}) \subset \overset{\circ}{\mathbb{K}}\ , \quad \text{for any} \quad t > 0\ , \quad \text{iff } A \text{ is irreducible}\ ,$$

Assume A is a quasimonotone matrix.

As a consequence of (i) and (ii) in Theorem A.31 we have the following.

If z_1, $z_2 \in \mathbb{K}$, and $z_1 \leq z_2$ we have $z_2 - z_1 \in \mathbb{K}$ so that $e^{At}\,(z_2 - z_1) \in \mathbb{K}$; hence

$$z_1\ ,\ z_2 \in \mathbb{K}\ ,\ z_1 \leq z_2 \implies e^{At}\,z_1 \leq e^{At}\,z_2\ , \qquad t \in \mathbb{R}_+\ . \tag{A.35}$$

We shall say that a quasimonotone (nonnegative) matrix A induces a "monotone flow" on system (A.32).

Further, if in addition A is irreducible, we have

$$z_1\ ,\ z_2 \in \mathbb{K}\ ,\ z_1 < z_2 \implies e^{At}\,z_1 \ll e^{At}\,z_2\ , \qquad t > 0\ . \tag{A.36}$$

i.e. A induces a "strongly monotone flow" [72, 139].

A.4.2. Nonlinear autonomous quasimonotone systems

Differential inequalities can be used to show that the flow generated by the differential system

$$\frac{dz}{dt} = f(z) \tag{A.37}$$

is monotone.

Theorem A.32. [128, 72] *Suppose $f \in C^1(\mathring{\mathbb{K}}\ ,\ \mathbb{R}^n)$, and that f is a quasimonotone (cooperative) vector function. If $\{z(t)\ ,\ t \in \mathbb{R}_+\}$ satisfies*

$$\frac{dz}{dt} \leq f(z)\ ; \tag{A.38}$$

if $\{y(t)\ ,\ t \in \mathbb{R}_+\}$ satisfies

$$\frac{dy}{dt} \geq f(y) \tag{A.39}$$

and if

$$z(0) = z_0 \leq y_0 = y(0) \tag{A.40}$$

then,

$$z(t) \leq y(t)\ , \qquad \text{for all} \quad t \in \mathbb{R}_+\ . \tag{A.41}$$

In particular if both $z(t; z_0)$ and $y(t; z_0)$ are solutions of the ODE system (A.37) but their initial conditions satisfy (A.40), then (A.41) still holds. We may again state that system (A.37) is order preserving, or that it generates a monotone flow.

Whenever f is such that its Jacobi matrix $J_f(z)$ at any $z \in \mathring{\mathbb{K}}$ is irreducible, then for any $z_0\ ,\ y_0 \in \mathbb{K}$

$$z_0 < y_0 \implies z(t) \ll y(t)\ , \qquad t > 0\ ,$$

i.e. system (A.37) generates a "strongly monotone flow".

In particular, if we assume for $f \in C^1(\mathring{\mathbb{K}}, \mathbb{R}^n)$ the following hypotheses

(F1) $f(0) = 0$

(F2) f is quasimonotone (cooperative) in $\mathbb{K}$

(F3) for any $\xi \in \mathbb{K}$ a $\xi_0 \in \mathbb{K}$ exists, $\xi_0 \gg 0$ such that $\xi \leq \xi_0$ and $f(\xi_0) \ll 0$, we may claim that for any choice of $z_0 \in \mathbb{K}$, a unique global solution of system (A.37) exists, subject to the initial condition z_0 at $t = 0$.

We may denote such a solution by $\{V(t)\, z_0\ ,\ t \in \mathbb{R}_+\}$ so that a (nonlinear) C_o-semigroup of evolution operators $\{V(t)\ ,\ t \in \mathbb{R}_+\}$ is defined for system (A.37) (see Appendix B; Section B.1.4).

In fact, thanks to the above mentioned comparison theorem we may state that the evolution operator satisfies the following properties.

(i) $V(0) = I$

(ii) $V(t+s) = V(t)\, V(s)\ , \qquad s\,,\, t \geq 0$

(iii) $V(t)\, 0 = 0\ , \qquad t \geq 0$

(iv) for any $t \geq 0$, the mapping $z_0 \in \mathbb{K} \longrightarrow V(t)\, z_0 \in \mathbb{R}^n$ is continuous, uniformly in $t \in [t_1, t_2] \subset \mathbb{R}_+$

(v) for any $z_0 \in \mathbb{K}$, the mapping $t \in \mathbb{R}_+ \longrightarrow V(t)\, z_0 \in \mathbb{K}$ is continuous

(vi) for any $z_1\,,\ z_2 \in \mathbb{K}$, $z_1 \leq z_2 \Longrightarrow V(t)\, z_1 \leq V(t)\, z_2$, for any $t \in \mathbb{R}_+$

(vii) for any $t \in \mathbb{R}_+\ :\ V(t)\, \mathbb{K} \subset \mathbb{K}$

The following lemma is a consequence of Theorem A.31 and of Theorem A.32.

Lemma A.33. *Under the assumptions (F1)-(F3) for f, if further a quasimonotone irreducible $n \times n$ matrix B exists for which a $\delta > 0$ exists such that*

$$\text{(A.42)} \qquad \textit{for any} \quad \xi \in \mathbb{K}\ ,\ |\xi| \leq \delta\ :\ f(\xi) \geq B\xi$$

then the evolution operator $V(t)$ of system (A.37) is strongly positive for any $t > 0$; i.e.

$$z_0 \in \mathbb{K}\ ,\ z_0 \neq 0 \quad \Longrightarrow \quad V(t)\, z_0 \gg 0\ , \quad t > 0$$

or

$$\text{(A.43)} \qquad V(t)\left(\dot{\mathbb{K}}\right) \subset \mathring{\mathbb{K}}\ , \quad t > 0$$

(we have denoted by $\dot{\mathbb{K}} := \mathbb{K} - \{0\}$).

Suppose now that the Jacobi matrix $J_f(z)$ of f be irreducible for any $z \in \mathbb{K}$, and nonincreasing in z, i.e.

$$0 \leq \eta \leq \xi \implies (J_f(\eta))_{ij} \geq (J_f(\xi))_{ij}\ , \qquad i,j = 1,\cdots,n\ .$$

This implies that

(F4) for any $R > 0$ there exists a quasimonotone positive irreducible matrix C_R such that

$$0 \leq \eta \leq \xi \,, \ \|\xi\| \,, \|\eta\| \leq R \implies f(\xi) - f(\eta) \geq C_R \, (\xi - \eta)$$

Lemma A.34. [161] *Under assumptions (F4) for f in system (A.37), if $z_0 \,,\, y_0 \in \mathbb{K}$,*

$$z_0 < y_0 \implies V(t)\, z_0 \ll V(t)\, y_0 \ , \qquad t > 0 \ .$$

Finally if we introduce the property of strict sublinearity for f

(F5) for any $z \in \mathbb{K}$ and for any $\tau \in (0,1)$

$$\tau\, f(z) < f(\tau z)$$

the following lemma holds.

Lemma A.35. [161] *Under assumptions (F4) and (F5) for f the evolution operator $V(t)$ of system (A.37) is strongly concave for any $t > 0$. This means that*

(A.44) *for any $z_0 \in \overset{\circ}{\mathbb{K}}$, and for any $\sigma \in (0,1)$ an $\alpha = \alpha(z_0, \sigma) > 0$ exists such that*

$$V(t)\,(\sigma z_0) \geq (1+\alpha)\,\sigma\, V(t)\, z_0 \ , \qquad t > 0$$

We anticipate here the following theorem (see Section B.1.1 for definitions):

Theorem A.36. [137] *Let E be a real Banach space with cone $\mathbb{K}$. If an operator A on E is strongly positive, strongly monotone and strongly concave with respect to $\mathbb{K}$, then A cannot have two distinct nontrivial fixed points in the cone $\mathbb{K}$.*

Remark. Clearly Theorem A.36 excludes the existence of more than one nontrivial equilibrium for system (A.37) under assumptions of Lemmas A.33-A.35, since in this case $V(t)$ is, for any $t > 0$, a strongly positive, strongly monotone and strongly concave operator, and any equilibrium of system (A.37) is a fixed point for any $V(t)$, $t > 0$.

A.4.2.1. Lower and upper solutions, invariant rectangles, contracting rectangles

We say that z^* is an equilibrium solution for system (A.37) if it satisfies the system

$$f(z) = 0 \tag{A.45}$$

We shall say that $\underline{z} \in \mathbb{K}$ is a "lower (upper) solution" for (A.45) if

$$f(\underline{z}) \geq 0 \qquad (\leq 0)\ . \tag{A.46}$$

Theorem A.37. (Invariant Rectangles) [39] *Under the assumptions of Theorem A.32, if $\underline{z}$, $\overline{z}$ are a lower solution, respectively an upper solution, of system (A.45), then*

$$\mathcal{R} := [\underline{z}, \overline{z}] = \{z \in \mathbb{K} \mid \underline{z} \leq z \leq \overline{z}\}$$

is an invariant rectangle for system (A.37).

Remark. Note that the existence of an invariant rectangle $\mathcal{R}$ for system (A.37) insures the global existence of solutions with initial condition in $\mathcal{R}$.

Theorem A.38. (Nested Invariant Rectangles) [39] *Let z^* be an equilibrium for (A.37) in $\mathring{\mathbb{K}}$, and let $\theta\mathcal{R}$ be the following rectangle in $\mathbb{K}$*

$$\theta\mathcal{R} := \{z \in \mathbb{K} \mid |z_i - z_i^*| \leq \theta a_i\ , \quad i = 1, \cdots, n\} \tag{A.47}$$

with $\theta \in [0,1]$, and $a_i \in \mathbb{R}_+$, $i = 1, \cdots, n$. If, for any $\theta \in [0,1]$, $\theta\mathcal{R}$ is an invariant rectangle, then z^* is stable as an equilibrium solution of system (A.37).

We shall say that a bounded rectangle $\mathcal{R} \subset \mathbb{K}$ is "contracting" for the vector field f if at every point $z \in \partial\mathcal{R}$ we have

$$f(z) \cdot \nu_z < 0 \tag{A.48}$$

where ν_z is the outward pointing normal at z.

Theorem A.39. (Nested Contracting Rectangles)[185, 39] *Under the assumptions of Theorem A.38, if further, for any $\theta \in (0,1]$, the rectangle (A.47) is contracting for the vector field f, then z^* is asymptotically stable as an equilibrium solution of system (A.37); globally in $\mathcal{R}$.*

Actually the proof of Theorem A.39 is based on the Lyapunov's direct method which will be introduced later.

A.5. Lyapunov methods, LaSalle Invariance Principle

In the previous section the concepts of stability and asymptotic stability of an equilibrium solution of an ODE system were introduced.

Actually, apart from the quasimonotone case, the only technical tool to show that stability holds was based on the knowledge of the eigenvalues of the Jacobi matrix of the function f at the equilibrium. But this provides only information on the local behavior at equilibrium in the case of nonlinear systems.

The methods due to Lyapunov provide a means for identifying the region of attraction of a critical point.

We shall assume that 0 is a critical point for our system. Concepts can be easily extended to any other point.

Let $\Omega \subset \mathbb{R}^n$ be an open set in $\mathbb{R}^n$ such that $0 \in \Omega$.

A scalar function

$$V : \Omega \longrightarrow \mathbb{R}$$

is "positive semidefinite" on Ω if it is continuous on Ω, $V(0) = 0$ and

$$V(z) \geq 0 \ , \qquad z \in \Omega \ .$$

A scalar function V is "positive definite" on Ω if it is positive semidefinite on Ω and

$$V(z) > 0 \ , \qquad z \in \Omega - \{0\} \ .$$

A scalar function V is "negative semidefinite (negative definite)" on Ω if $-V$ is positive semidefinite (positive definite) on Ω.

Lemma A.40. (Sylvester) [27] *The quadratic form*

$$z^T A z = \sum_{i,j=1}^{n} a_{ij} z_i z_j \ , \qquad z \in \mathbb{R}^n$$

associated with the $n \times n$ symmetric matrix $A = A^T$ is positive definite iff

$$det\,(a_{ij} \ ; \ i,j = 1, \cdots, s) > 0$$

for any $s = 1, \cdots, n$.

Consider the differential equation

$$\frac{dz}{dt} = f(z) \tag{A.49}$$

where $f \in C\,(D\,, \mathbb{R}^n)$ satisfies enough smoothness properties to ensure that a solution of (A.49) exists through any point in D, is unique and depends continuously upon the initial data (we shall assume that $0 \in D$ and that $f(0) = 0$ so that 0 is an equilibrium solution of (A.49)).

Let $\Omega \subset D$ be an open subset of D in $\mathbb{R}^n$ and let $V \in C^1(\Omega, \mathbb{R})$. We define $\dot{V}$ with respect to system (A.49) as

$$\dot{V}(z) = grad\, V(z) \cdot f(z) \ , \qquad z \in \Omega \ . \tag{A.50}$$

If $z(t)$ is a solution of (A.49), then the total derivative of $V(z(t))$ with respect to $t \in \mathbb{R}_+$ is

$$\frac{d}{dt} V(z(t)) = \dot{V}(z(t)) \ ;$$

that is $\dot{V}$ is the derivative of V along the trajectories of (A.49).

Theorem A.41. (Lyapunov) [104] *If there is a positive definite function $V \in C^1(\Omega, \mathbb{R})$ on the open subset $\Omega \subset D$, such that $0 \in \Omega$, with $\dot{V}$ negative semidefinite in Ω, then the solution $z = 0$ is stable for system (A.49). If, in addition, $\dot{V}$ is negative definite on Ω, then the solution $z = 0$ is asymptotically stable for system (A.49), globally in Ω.*

Lemma A.42. (Lyapunov) [104, 38] *Let A be an $n \times n$ real matrix. The matrix equation*

$$A^T B + BA = -C \tag{A.51}$$

has a positive definite solution B (which is symmetric), for every positive definite symmetric matrix C, iff A is a stable matrix, i.e. $\mathcal{R}e\, \lambda < 0$ for any $\lambda \in \sigma(A)$.

As an application of Lemma A.42, consider the linear differential system

$$\frac{dz}{dt} = Az \tag{A.52}$$

and the scalar function

$$V(z) = z^T B\, z \qquad z \in \mathbb{R}^n$$

where B is a positive definite symmetric matrix then, with respect to (A.52)

$$\dot{V}(z) = z^T (A^T B + BA)\, z \ , \qquad z \in \mathbb{R}^n$$

According to Lemma A.42, if A is stable we may choose B so that $A^T B + BA$ is negative definite.

Thus the stability of A implies the (global) asymptotic stability of 0 for system (A.52), as already known by direct argument.

Let $V \in C^1(\Omega, \mathbb{R})$ be a positive definite function on the open set $\Omega \subset D$ such that $0 \in D$. We say that V is a "Lyapunov function" for system (A.49) if

$$\dot{V}(z) = grad\, V(z) \cdot f(z) \leq 0 \;, \qquad \text{in} \quad \Omega \;. \tag{A.53}$$

Theorem A.43. (LaSalle Invariance Principle)[104, 216] *Let V be a Lyapunov function for system (A.49) in an open subset $\Omega \subset D$, and let V be continuous on $\overline{\Omega}$, the closure of Ω. Let*

$$E := \left\{ z \in \overline{\Omega} \mid \dot{V}(z) = 0 \right\}$$

and let M be the largest invariant subset of (A.49) in E. Suppose that for any initial point $z_0 \in \Omega$ the positive orbit $\Gamma_+(z_0)$ of (A.49) lies in Ω and is bounded. Then the ω-limit set of $\Gamma_+(z_0)$, $\omega\left(\Gamma_+(z_0)\right) \subset M$, so that

$$\lim_{t \to +\infty} dist\left(z(t; z_0)\;,\; M\right) = 0$$

Corollary A.44. *Under the same assumptions as in Theorem A.43 if $M = \{z^*\}$, with $f(z^*) = 0$, then the equilibrium solution z^* is a global attractor in Ω, for system (A.49).*

Corollary A.45. [104] *If V is a Lyapunov function on*

$$\Omega = \{z \in \mathbb{R}^n \mid V(z) < \rho\}$$

and Ω is bounded, then every solution of (A.49) with initial value in Ω approaches M as $t \to +\infty$.

Corollary A.46. [104] *If V is a Lyapunov function in $\mathbb{R}^n$, bounded from below, and such that $V(z) \;\to +\infty$ as $\|z\| \;\to +\infty$, then every positive orbit of (A.49) is bounded and approaches the largest invariant subset M of $E' := \{z \in \mathbb{R}^n \mid \dot{V}(z) = 0\}$, as $t \to +\infty$. In particular if $M = \{z^*\}$, with $f(z^*) = 0$, then the equilibrium solution z^* is globally asymptotically stable in $\mathbb{R}^n$ for system (A.49).*

Appendix B. Dynamical systems in infinite dimensional spaces

B.1. Banach spaces

An abstract "linear space" X over $\mathbb{R}$ (or $\mathbf{C}$; unless explicitly stated we shall always refer to $\mathbb{R}$) is a collection of elements such that for each $z_1,\ z_2 \in X$ the sum $z_1 + z_2 \in X$ is defined such that

(i) $z_1 + z_2 = z_2 + z_1$

and an element $0 \in X$ exists such that

(ii) $0 + z = z + 0 = z \quad$ for all $\quad z \in X$.

Also, for any number $a \in \mathbb{R}$ and any element $z \in X$ the scalar multiplication is defined $az \in X$ such that

(iii) $1\, z = z$, $\quad$ for all $\quad z \in X$

(iv) $a(bz) = (ab)z = b(az)$, $\quad$ for all $\quad a, b \in \mathbb{R} \quad$ and all $\quad z \in X$

(v) $(a + b)z = az + bz$, $\quad$ for all $\quad a, b \in \mathbb{R} \quad$ and all $\quad z \in X$.

A linear space X is a "normed linear space" if to each $z \in X$ there corresponds a nonnegative real number $\|z\|$ called the "norm" of z which satisfies

(a) $\|z\| = 0 \iff z = 0$

(b) $\|z_1 + z_2\| \leq \|z_1\| + \|z_2\| \quad$ for all $\quad z_1,\ z_2 \in X$

(c) $\|az\| = |a|\ \|z\| \quad$ for all $\quad a \in \mathbb{R} \quad$ and all $\quad z \in X$

When confusion may arise, we will write $\|\cdot\|_X$ for the norm on X.
A norm on a linear space X induces a metric via the following distance

$$dist(z_1, z_2) = \|z_1 - z_2\| \ , \quad \text{for all} \quad z_1\ ,\ z_2 \in X\ .$$

It is such that

(d1) $dist(z_1, z_2) = 0 \iff z_1 = z_2$

(d2) $dist(z_1, z_2) = dist(z_2, z_1)$

(d3) $dist(z_1, z_3) \leq dist(z_1, z_2) + dist(z_2, z_3)$

for all z_1, z_2, $z_3 \in X$.

X endowed with $dist$ is a "metric space".

A sequence $(z_n)_{n \in \mathbb{N}}$ in a normed linear space X converges in X if a $z \in X$ exists such that $\lim\limits_{n \to \infty} \|z_n - z\| = 0$.

A sequence $(z_n)_{n \in \mathbb{N}}$ is a Cauchy sequence in X if for every $\epsilon > 0$ a $\nu \in \mathbb{N}$ exists such that for any $n, m \in \mathbb{N}$ $n, m > \nu$: $\|z_n - z_m\| < \epsilon$.

The space X is "complete" if every Cauchy sequence in X is convergent in X.

A "Banach space" X is a complete normed linear space.

For $z \in X$, a normed linear space, the "open ball" about $z_0 \in X$ with radius $\rho > 0$ is the set $B_\rho(z_0) := \{z \in X \mid \|z - z_0\| < \rho\}$. If $A \subset X$ then z is an "interior point" of A if a $\rho > 0$ exists so that $B_\rho(z) \subset A$.

The set of all interior points of A is called the "interior" of A and is usually denoted by $\overset{\circ}{A}$. Clearly $\overset{\circ}{A} \subset A$. A set $A \subset X$ is "open" if $A = \overset{\circ}{A}$.

A set $A \subset X$ is "bounded" if for any $z \in A$ a $\rho > 0$ exists such that $A \subset B_\rho(z)$.

We say that $z \in X$ is a "limit point" of A if a sequence of elements of $A \subset X$ exists such that $z = \lim\limits_{n \to \infty} z_n$. The "closure" $\overline{A}$ of A is the set of all limit points of A. Clearly $A \subset \overline{A}$. A set A is "closed" if $A = \overline{A}$. $A \subset X$ is "dense" in X if $\overline{A} = X$.

Let X be a complete metric space and let $A \subset X$. If $A = \overline{A}$ then A is a complete subspace of X.

A is "precompact" in X if every sequence of elements of A contains a Cauchy subsequence; A is "compact" if every sequence of elements of A contains a subsequence convergent to a point in A.

In a metric space X we say that $A \subset X$ is "relatively compact" if $\overline{A}$ is compact. A subset $A \subset X$ which is relatively compact is also precompact. In a complete metric space X a precompact subset of X is also relatively compact, and thus the two concepts coincide.

In a complete metric space A is compact iff A is precompact and closed.

In a metric space X, every subset $A \subset X$ which is compact is also closed. Every precompact subset A is bounded.

A normed linear space X is "locally compact" if every closed and bounded subset is compact.

Consider $\mathbb{R}^n$ equipped with the Euclidean norm. Then $\mathbb{R}^n$ is complete and locally compact.

A vector space X is said to be "finite dimensional" if there is a positive integer $n \in \mathbb{N}$ such that X contains a set of n linearly independent vectors whereas any set of $n+1$ or more vectors of X is linearly dependent. This number n is called "dimension" of the space X.

If X is not finite dimensional then it is said to be "infinite dimensional".

If $dim X = n$ any set of n linearly independent vectors of X is called a "basis" for X.

If $\{e_1, \cdots, e_n\}$ is a basis for X then any $x \in X$ has a unique representation as a linear combination of the basis vectors.

If X is a finite dimensional normed space

$$M \text{ compact} \iff M \text{ closed and bounded.}$$

Thus every finite dimensional normed space is "locally compact".

The converse is also true.

Theorem B.1. (F. Riesz) [140] *A locally compact linear normed space has finite dimension.*

Every finite dimensional normed space is complete.

A norm $\|\cdot\|$ on a vector space X is said to be "equivalent" to another norm $\|\cdot\|_o$ on X if there are positive real numbers a and b such that for all $x \in X$ we have

$$a \|x\|_o \leq \|x\| \leq b \|x\|_o$$

On a finite dimensional vector space all norms are equivalent.

$\mathbb{R}^n$ equipped with the Euclidean norm, is a finite dimensional Banach space.

$C([0,1])$, the set of all continuous real-valued functions defined on the closed interval $[0,1] \subset \mathbb{R}$ is a real Banach space, when equipped with the norm

$$\|u\| = \max_{x \in [0,1]} |u(x)| \ , \quad \text{for } u \in C([0,1]) \ .$$

$C^k([0,1])$, the set of continuous real-valued functions having $k \in \mathbb{N}$ continuous derivatives on $[0,1] \subset \mathbb{R}$, with norm

$$\|u\|_k = \sum_{s=0}^{k} \max_{x \in [0,1]} \left| u^{(s)}(x) \right|$$

is a real Banach space ($u^{(s)}$ denotes the derivative of order s of the function u).

B.1.1. Ordered Banach spaces

Let E be a real vector space.

An ordering $\leq$ on E is called linear if

(i) $x \leq y \Longrightarrow x+z \leq y+z \quad$ for all $\quad x,\, y,\, z \in E$

(ii) $x \leq y \Longrightarrow \alpha x \leq \alpha y \quad$ for all $\quad x,\, y \in E\,,\ \alpha \in \mathbb{R}_+$.

A real vector space together with a linear ordering is called an "ordered vector space (OVS)" [0].

Let V be an OVS and let $P := \{x \in V \mid x \geq 0\}$. Clearly P has the following properties

(P1) $P + P \subset P$

(P2) $\mathbb{R}_+ P \subset P$

(P3) $P \cap (-P) = \{0\}$

A nonempty subset P of a real vector space V satisfying (P1-P3) is called a "cone".

A cone P is called "generating" if $E = P - P$.

Every cone P in a real vector space E induces a partial linear ordering on E by

$$x \leq_P y \quad \overset{\text{def}}{\Longleftrightarrow} \quad y - x \in P\ .$$

The elements in $\dot{P} := P - \{0\}$ are called "positive" and P is called the "positive cone" of the ordering.

Consequently for every linear space E there is a one-to-one correspondence between the family of linear orderings and the family of cones.

A set A is said "order convex" whenever $x,\, y \in A$ implies $[x, y] \subset A$, where

$$[x, y] := \{z \in E \mid x \leq z \leq y\}\ .$$

A cone is order convex.

Let $E = (E, \|\cdot\|)$ be a Banach space ordered by a cone P. Then E is called an "ordered Banach space (OBS)" if the positive cone is closed.

Proposition B.2. *Let E be an ordered Banach space with respect to a cone P. If $\overset{\circ}{P} \neq \emptyset$ then P is generating.*

The Euclidean space $\mathbb{R}^n$ is an ordered Banach space with respect to the cone $\mathbb{K} := \mathbb{R}^n_+ := \{z \in \mathbb{R}^n \mid z_i \geq 0\,,\ i = 1, \cdots, n\}$. This cone has a nonempty interior $\overset{\circ}{\mathbb{K}} = \mathbb{R}^{n*}_+ := \{z \in \mathbb{R}^n \mid z_i > 0\,,\ i = 1, \cdots, n\}$, hence it is generating.

Let Ω be a nonempty bounded open subset of $\mathbb{R}^m$, $m \geq 1$. For any $k \in \mathbb{N}$, we denote by $C^k(\overline{\Omega})$ the vector space of all uniformly continuous functions $u : \Omega \longrightarrow \mathbb{R}$ such that all the partial derivatives of order up to k exist and are uniformly continuous on Ω.

Due to the uniform continuity, each of the derivatives $D^\alpha u := D_1^{\alpha_1} \cdots\ D_m^{\alpha_m} u\,,\ \ \alpha = (\alpha_1, \cdots, \alpha_m) \in \mathbb{N}^m\,,\ \ |\alpha| = \sum_{i=1}^m \alpha_i \leq k$, has a unique continuous extension over $\overline{\Omega}$.

We define

$$\|u\|_k := \sum_{|\alpha| \leq k} \max_{x \in \overline{\Omega}} |D^\alpha u(x)|\ .$$

Equipped with this norm, each $C^k(\overline{\Omega})$ is a Banach space. With the ordering induced by $C_+(\overline{\Omega}) := \{u \in C(\overline{\Omega}) \mid u(x) \geq 0\,,\ x \in \overline{\Omega}\}$, it is an ordered Banach space.

B.1.2. Functions

Given two sets X and Y, a function F from X to Y is a rule which assigns to any element x of a subset $\mathcal{D}(F) \subset X$ a unique $y \in Y$ that we denote by $F(y)$. This is denoted by

$$F : (\mathcal{D}(F) \subset X) \longrightarrow Y\ .$$

$\mathcal{D}(F)$ is called the "domain" of F and $\mathcal{R}(F) := \{y \in Y \mid y = F(x)\,,\ \text{for some } x \in \mathcal{D}(F)\}$ is the "range" of F.

If $\mathcal{R}(F) = Y$ the function is "onto". If $F(x) = F(x') \implies x = x'$, then F is "one-to-one"; in this case there exists an "inverse function" $F^{-1} : (\mathcal{R}(F) \subset Y) \longrightarrow X$ such that $F^{-1}F(x) = x$ for all $x \in \mathcal{D}(F) = \mathcal{R}(F^{-1})$ (and $FF^{-1}(y) = y$ for all $y \in \mathcal{D}(F^{-1}) = \mathcal{R}(F)$).

The set $G_F := \{(x, y) \in X \times Y \mid x \in \mathcal{D}(F)\,,\ y = F(x)\}$ is the "graph" of F.

The identity on X is the function $I : X \longrightarrow X$ defined for all $x \in X$: $I(x) = x$.

If X is a linear space on $\mathbb{R}$ a function $F : (\mathcal{D}(F) \subset X) \longrightarrow \mathbb{R}$ is also called a (real) functional on X.

Let X, Y be linear spaces on $\mathbb{R}$. A function $F : (\mathcal{D}(F) \subset X) \longrightarrow Y$ is a "linear operator" on X if $\mathcal{D}(F)$ is a linear subspace of X and

$$F(\alpha x + \beta x') = \alpha F(x) + \beta F(x')$$

for any $x, x' \in \mathcal{D}(F)$, $\alpha, \beta \in \mathbb{R}$ (for a linear function F we shall usually write Fx for $F(x)$).

The "null space", or "kernel" of a linear operator F is the set

$$ker(F) := \{x \in \mathcal{D}(F) \mid Fx = 0\} .$$

Clearly $0 \in ker(F)$.

If $F : (\mathcal{D}(F) \subset X) \longrightarrow Y$ is a linear operator between two real linear spaces X and Y , then $\mathcal{R}(F)$ is a vector space. If $dim\mathcal{D}(F) = n < +\infty$, then $dim\mathcal{R}(F) \leq n$. The null space $ker(F)$ is a vector space.

The inverse of a linear operator $F : (\mathcal{D}(F) \subset X) \longrightarrow Y$ exists iff $ker(F) = \{0\}$. If F^{-1} exists it is itself a linear operator. If $dim\mathcal{D}(F) = n < +\infty$, and F^{-1} exists, then $dim\mathcal{R}(F) = dim\mathcal{D}(F)$.

Let $F : (\mathcal{D}(F) \subset X) \longrightarrow Y$, X, Y metric spaces. F is "continuous" at $x_0 \in \mathcal{D}(F)$ if for every $\epsilon > 0$ there exists a $\delta(x_0, \epsilon) > 0$ such that $dist_X(x, x_0) < \delta \Longrightarrow dist_Y(F(x), F(x_0)) < \epsilon$. Equivalently, if for every sequence $(x_n)_{n\in\mathbb{N}} \subset \mathcal{D}(F)$, converging to $x_0 \in \mathcal{D}(F)$, the sequence $(F(x_n))_{n\in\mathbb{N}} \subset Y$ converges to $F(x_0)$.

A function F is said to be "continuous" if it is continuous at every $x_0 \in \mathcal{D}(F)$. It is said "uniformly continuous" if, in $\mathcal{D}(F)$, $\delta(\epsilon, x_0)$ can be chosen independently of $x_0 \in \mathcal{D}(F)$.

If S is a compact subset of $\mathcal{D}(F)$ and F is continuous on S then F is uniformly continuous on S.

A function $F : (\mathcal{D}(F) \subset X) \longrightarrow X$, on a metric space X is a "contraction" if $\mathcal{R}(F) \subset \mathcal{D}(F)$ and , for some real $0 \leq \alpha < 1$: $dist(F(x), F(x')) \leq \alpha\, dist(x, x')$ for all x, $x' \in \mathcal{D}(F)$.

Theorem B.3. (Banach-Caccioppoli)[104] *Let $F : X \longrightarrow X$ be a contraction, X a complete metric space. Then there exists a unique "fixed point" $x_0 \in X$ such that*

$$F(x_0) = x_0 .$$

A contraction on a metric space X is uniformly continuous on X.

A function $F : (\mathcal{D}(F) \subset X) \longrightarrow Y$, X, Y normed linear spaces, is "Lipschitz continuous" at $x_0 \in \mathcal{D}(F)$ if a real $\alpha(x_0) > 0$ exists such that $\|F(x) - F(x_0)\|_Y \leq \alpha \|x - x_0\|_X$ for all $x \in B_\rho(x_0)$, for some $\rho > 0$.

A function F is said to be Lipschitz continuous on $S \subset \mathcal{D}(F)$ if it is Lipschitz continuous at each $x_0 \in S$.

F is "uniformly Lipschitz continuous on S"$\subset \mathcal{D}(F)$ if $\alpha(x_0)$ can be chosen independent of $x_0 \in S$.

If S is precompact, Lipschitz continuity on S implies uniform Lipschitz continuity on S.

A contraction on a normed linear space X is uniformly Lipschitz continuous on X.

If $F : X \longrightarrow Y$ is one-to-one and continuous together with its inverse then we say that F is an "homeomorphism" of X onto Y.

Theorem B.4. (Brower)[104] *Any continuous function of the closed unit ball in* $\mathbb{R}^n$ *into itself must have a fixed point.*

Corollary B.5. [104] *If* $A \subset \mathbb{R}^n$ *is homeomorphic to the closed unit ball in* $\mathbb{R}^n$ *and* f *is continuous from* A *into* A, *then* f *has a fixed point in* A.

A subset A of a Banach space is "convex" if for any x, $y \in A$ it follows that $tx + (1-t)y \in A$ for any $t \in [0,1]$.

Theorem B.6. (Schauder) [104] *If* A *is a convex, compact subset of a Banach space* X *and* $f : A \longrightarrow A$ *is continuous, then* f *has a fixed point in* A.

A function $F : (\mathcal{D}(F) \subset X) \longrightarrow Y$, X and Y metric spaces, is "bounded" (resp. "compact") if F maps bounded sets in $\mathcal{D}(F)$ into bounded sets (resp. precompact sets) in Y.

If X and Y are Banach spaces F is compact iff for every bounded set $A \subset \mathcal{D}(F)$, $\overline{F(A)}$ is compact in Y. If in addition F is continuous it is called "completely continuous".

Corollary B.7. [104] *If* A *is a closed, convex, bounded subset of a Banach space* X *and* $f : A \longrightarrow A$ *is completely continuous, then* F *has a fixed point in* A.

If X, Y are metric spaces, $F : (\mathcal{D}(F) \subset X) \longrightarrow Y$, $\mathcal{D}(F)$ is compact and F is continuous on $\mathcal{D}(F)$, then F is uniformly continuous and $\mathcal{R}(F)$ is compact in Y.

Let $F : (\mathcal{D}(F) \subset X) \longrightarrow Y$ be linear, X, Y normed linear spaces. Then F is bounded iff

$$\sup_{0 \neq x \in \mathcal{D}(F)} \frac{\|Fx\|_Y}{\|x\|_X} = \alpha < +\infty \ .$$

In this case the quantity α will be called the "norm" of F and shall be denoted by $\|F\|$.

Clearly, for any $x \in \mathcal{D}(F)$:

$$\|Fx\|_Y \leq \|F\| \ \|x\|_X \ .$$

A real $n \times n$ matrix A is a bounded linear operator on $\mathbb{R}^n$.

If a normed linear space X is finite dimensional, then every linear operator on X is bounded.

Let $F : (\mathcal{D}(F) \subset X) \longrightarrow Y$ be a linear operator, X, Y normed linear spaces. Then F is continuous iff F is bounded. If F is continuous at a point $x_0 \in \mathcal{D}(F)$, then F is continuous, uniformly continuous and uniformly Lipschitz continuous on $\mathcal{D}(F)$.

If F is a bounded linear operator then $ker(F)$ is closed in X.

Let $F : (\mathcal{D}(F) \subset X) \longrightarrow Y$, X, Y complete metric spaces. We say that F is a "closed operator" if its graph is a closed set in $X \times Y$ equipped with the product metric

$$dist_{X\times Y}((x,y)\,,\,(x',y')) := dist_X(x,x') + dist_Y(y,y')\ .$$

Equivalently if given a Cauchy sequence $(x_n)_{n\in\mathbb{N}}$ in $\mathcal{D}(F) \subset X$ such that $(Fx_n)_{n\in\mathbb{N}}$ is Cauchy in Y, then $x_n \longrightarrow x_0 \in \mathcal{D}(F)$ and $Fx_n \longrightarrow Fx_0$ in Y.

B.1.3. Linear operators on Banach spaces

Let now X be a complex Banach space.

For a linear operator $F : (\mathcal{D}(F) \subset X) \longrightarrow X$, the resolvent set $\rho(F)$ consists of all those $\lambda \in \mathbf{C}$ such that

(R1) $\lambda I - F$ is one-to-one

(R2) $\mathcal{R}(\lambda I - F)$ is dense in X

(R3) $(\lambda I - F)^{-1}$ is bounded.

Because of (R2) and (R3), for $\lambda \in \rho(F)$, there exists a unique extension $R(\lambda, F)$ of $(\lambda I - F)^{-1}$ to all of X; we shall call this extension the "resolvent" of F at $\lambda \in \rho(F)$.

We notice that if $\lambda \in \rho(F)$ then F is a closed operator iff $\mathcal{R}(\lambda I - F) = X$, since $(\lambda I - F)^{-1}$ is then closed bounded and densely defined. In this case $R(\lambda, F) = (\lambda I - F)^{-1}$.

The "spectrum" of F is $\sigma(F) := \{\lambda \in \mathbf{C} \mid \lambda \notin \rho(F)\}$.. If is seen that the spectrum may be partitioned as

$$\sigma(F) = \sigma_p(F) \cup \sigma_c(F) \cup \sigma_r(F)\ .$$

The "point spectrum" $\sigma_p(F)$ consists of those $\lambda \in \sigma(F)$ such that $(\lambda I - F)$ is not one-to-one; i.e. $(\lambda I - F)g = 0$ for some nonzero $g \in \mathcal{D}(F)$; then we say that λ is an "eigenvalue" of F and g is a corresponding "eigenvector".

The space $ker(\lambda I - F)$ is called the eigenspace of λ and its dimension is known as the "geometric multiplicity" of λ. The "generalized eigenspace" of λ, $\mathcal{M}(\lambda I - F)$ is the smallest closed linear subspace of X that contains $ker((\lambda I - F)^j)$ for $j \in \mathbb{N}^*$; its dimension is called the "algebraic multiplicity" of λ. Evidently the geometric multiplicity of λ is at most equal to its algebraic multiplicity. F is called "semisimple" if these two multiplicities agree for all $\lambda \in \sigma_p(F)$. An eigenvalue is "simple" if its algebraic multiplicity equals 1.

The "continuous spectrum" $\sigma_c(F)$ is the set of those $\lambda \in \sigma(F)$ for which $ker(\lambda I - F) = \{0\}$ and $\mathcal{R}(\lambda I - F)$ is dense in X but $R(\lambda, F)$ is unbounded.

Finally the "residual spectrum" $\sigma_r(F)$ is the set of those $\lambda \in \sigma(F)$ for which $ker(\lambda I - F) = \{0\}$ but $\mathcal{R}(\lambda I - F)$ is not dense in X.

Note that the spectrum of a linear operator on a finite dimensional space is a pure point spectrum , so that the spectrum is made of only eigenvalues.

If $X \neq \{0\}$ is a complex Banach space and F is a linear bounded operator on X, then $\sigma(F) \neq 0$.

The spectrum $\sigma(F)$ of a bounded linear operator on a complex Banach space X is compact and lies in the disk given by

$$|\lambda| \leq \|F\| \ .$$

The "spectral radius" $r(F)$ of a bounded linear operator F on a complex Banach space is the quantity

$$r(F) := \sup_{\lambda \in \sigma(F)} |\lambda| \ .$$

For a bounded linear operator F we have

$$r(F) \leq \|F\|$$

It can be shown further that

$$r(F) = \lim_{n \to \infty} \|F^n\|^{1/n}$$

Let X, Y be normed linear spaces and $F : (\mathcal{D}(F) \subset X) \longrightarrow Y$ a linear operator. Then if F is bounded and $dim\mathcal{R}(F) < +\infty$, the operator F is compact. If $dim(X) < +\infty$, the operator F is compact.

Theorem B.8. [140] *A compact linear operator F on a normed linear space X has the following properties.*

a) *Every spectral value $\lambda \neq 0$ is an eigenvalue. If $dimX = +\infty$ then $0 \in \sigma(F)$ (but may belong to either $\sigma_p(F)$ or $\sigma_c(F)$ or even $\sigma_r(F)$).*

b) *The set of all eigenvalues of F is countable. $\lambda = 0$ is the only possible point of accumulation of that set.*

c) *For $\lambda \neq 0$ the dimension of any eigenspace of F is finite.*

Let X, Y be OBS's with positive cones P and Q respectively.

A linear operator $F : X \longrightarrow Y$ is called "positive" if $F(P) \subset Q$, and "strictly positive" if $F(\dot{P}) \subset \dot{Q}$.

If Y is an OBS with respect to Q, and Q has nonempty interior, then F is called "strongly positive" if $F(\dot{P}) \subset \mathring{Q}$.

The following extends Perron-Frobenius theorem to the infinite dimensional case.

Theorem B.9. (Krein-Rutman) [1] *Let E be an OBS whose positive cone P has nonempty interior. Let F be a strongly positive compact linear operator on E. Then the following is true :*

(i) *The spectral radius $r(F)$ is positive.*

(ii) *$r(F)$ is a simple eigenvalue of F having a positive eigenvector and there is no other eigenvalue with a positive eigenvector.*

(iii) *for every $y \in \dot{P}$ the equation*

$$(\lambda I - F)x = y$$

has exactly one positive solution x if $\lambda > r(F)$, and no positive solution for $\lambda \leq r(F)$.

B.1.4. Dynamical systems and C_o-semigroups

For X a Banach space, a family of continuous operators $\{S(t)\,,\ t \in \mathbb{R}_+\}$ on X is a strongly continuous semigroup of continuous operators if

(i) $S(0) = I$

(ii) $S(t+\tau) = S(t)\,S(\tau)$ for all $t\,,\ \tau \in \mathbb{R}_+$

(iii) $S(\cdot)\,x \,:\, \mathbb{R}_+ \longrightarrow X$ is continuous for any $x \in X$.

Such a family is usually called a C_o-semigroup ((i) and (ii) define a semigroup; C_o refers to (iii)).

Clearly, every C_o-semigroup determines a dynamical system in X, and conversely, by the definition

$$u(t;x) := S(t)\,x \quad , \quad t \in \mathbb{R}_+ \quad , \quad x \in X \ .$$

Hence these two concepts are equivalent [215].

A dynamical system $\{S(t)\,,\ t \in \mathbb{R}_+\}$ is linear if $S(t)$ is a linear bounded operator on X for every $t \in \mathbb{R}_+$.

If $\{S(t)\,,\ t \in \mathbb{R}_+\}$ is a linear dynamical system on a Banach space X, there exist numbers $M \geq 1$, $\omega \in \mathbb{R}$ such that

$$\|S(t)\| \leq M\,e^{\omega t} \quad , \qquad t \in \mathbb{R}_+$$

With each linear dynamical system $\{S(t)\,,\ t \in \mathbb{R}_+\}$ on a Banach space X there is associated a certain linear operator $A \,:\, (\mathcal{D}(A) \subset X) \longrightarrow X$ called its "infinitesimal generator" defined as follows

$$\mathcal{D}(A) := \left\{ x \in X \;\middle|\; \text{there exists the limit} \quad \lim_{t \to 0^+} \frac{1}{t}\,[S(t)\,x - x] \right\}$$

$$Ax := \lim_{t \to 0^+} \frac{1}{t}\,[S(t)\,x - x] \ .$$

Theorem B.10. [179, 26] *Let $A \,:\, (\mathcal{D}(A) \subset X) \longrightarrow X$, X a Banach space, be the infinitesimal generator of a linear dynamical system $\{S(t)\,,\ t \in \mathbb{R}_+\}$. Then*

(i) A is a closed linear operator and $\mathcal{D}(A)$ is dense

(ii) $\mathcal{D}(A)$ *is positive invariant and for every* $x \in \mathcal{D}(A)$

$$\frac{d}{dt} S(t)\, x = A\, S(t)\, x = S(t)\, A\, x \;, \qquad t \in \mathbb{R}_+ \;.$$

Theorem B.11. (Hille-Phillips) [129, 26] *A linear operator* $A : (\mathcal{D}(A) \subset X) \longrightarrow X$ *on a complex Banach space* X *, is the infinitesimal generator of a linear dynamical system* $\{S(t) \;,\; t \in \mathbb{R}_+\}$ *satisfying* $\|S(t)\| \le M\, e^{\omega t}$ *for all* $t \in \mathbb{R}_+$ *, iff*

(i) A *is closed and* $\mathcal{D}(A)$ *is dense*

(ii) every real $\mu > \omega$ *is in the resolvent set* $\rho(A)$

(iii) $\|R(\mu, A)^n\| \le \dfrac{M}{(\mu - \omega)^n}$ *for all* $\mu > \omega, \;\; n = 1, 2, \cdots$.

Notice that condition (iii) is "easy" only if $\|R(\mu, A)\| \le \dfrac{1}{(\mu - \omega)}$ for all $\mu > \omega$.

Here $M > 0$, and $\omega \in \mathbb{R}$.

Let $\{S(t) \;,\; t \in \mathbb{R}_+\}$ be a linear dynamical system on a complex Banach space with infinitesimal generator $A : (\mathcal{D}(A) \subset X) \longrightarrow X$.

The following necessary conditions can be stated

Theorem B.12. [215] *If the equilibrium* $x^* = 0$ *is stable, then no eigenvalue of* A *has positive real part. If* $x^* = 0$ *is asymptotically stable, then every eigenvalue has negative real part.*

Remark. The above theorem cannot be reversed into sufficient conditions for stability and asymptotic stability in the infinite dimensional case.

On the other hand for linear systems on a Banach space the following statements are equivalent [215]:

(i) the equilibrium $x^* = 0$ is stable

(ii) every motion is stable

(iii) every positive orbit is bounded.

Further, the following statements are equivalent [215]:

(i) the equilibrium $x^* = 0$ is asymptotically stable

(ii) every motion is GAS.

For nonlinear dynamical systems such coincidences are not generally true.

The next step is to explore the behavior of a linear C_o-semigroup $\{S(t)\ ,\ t \in \mathbb{R}_+\}$ for large t that can be anyway drawn from the knowledge of the spectrum of its infinitesimal generator $A : (\mathcal{D}(A) \subset X) \longrightarrow X$.

Theorem B.13. [168] *Under the above assumptions, for any* $t \in \mathbb{R}_+$

(i) $e^{t\sigma(A)} \subset \sigma(S(t))$

(ii) $e^{t\sigma_p(A)} \subset \sigma_p(S(t)) \subset \left(e^{t\sigma_p(A)} \cup \{0\}\right)$.

Theorem B.14. [168] *If we define*

$$\omega_0 := \inf_{t>0} \frac{1}{t} \ln \|S(t)\|$$

we have

(i) $\lim_{t\to+\infty} \frac{1}{t} \ln \|S(t)\| = \omega_0$

(ii) for any $\omega > \omega_0$ *an* $M(\omega) > 0$ *exists such that* $\|S(t)\| \leq M(\omega)\, e^{\omega t}$, $t \in \mathbb{R}_+$.

(iii) $r(S(t)) = e^{\omega_o t}\ ,\quad t \in \mathbb{R}_+$.

The quantity $\omega_0(S)$ is usually called the "growth bound" of the C_o-semigroup $\{S(t)\ ,\ t \in \mathbb{R}_+\}$. For compact semigroups we may relate ω_0 to $s(A) := \sup\{\mathcal{R}e\ \lambda \mid \lambda \in \sigma(A)\}$ as a consequence of Theorem B.8 p.248.

Theorem B.15. [168, 215] *Suppose that for some* $t_0 > 0$, $S(t_0)$ *is compact. Then*

$$\omega_0 = s(A)\ .$$

Theorem B.16. [215] *Under the above assumptions if the resolvent* $R(\mu, A) = (I - \mu A)^{-1}$ *is compact for some* $\mu \in (0, \lambda_0)$, $\lambda_0 > 0$ *then every bounded orbit is precompact.*

B.2. The initial value problem for systems of semilinear parabolic equations (reaction-diffusion systems)

We shall consider now dynamical systems in Banach spaces defined by systems of semilinear parabolic equations of the form

$$(B.1) \qquad \frac{\partial}{\partial t}u = D\Delta u + f(u) \ , \qquad \text{in } \Omega \times \mathbb{R}_+$$

subject to boundary conditions

$$(B.1b) \qquad \beta_i \frac{\partial}{\partial \nu}u_i + \alpha_i u_i = 0 \ , \qquad \text{in } \partial\Omega \times \mathbb{R}_+ \ ,$$

for $i = 1, \cdots, n$, and initial conditions

$$(B.1o) \qquad u(0) = u_0 \ , \qquad \text{in } \overline{\Omega} \ .$$

(here $\partial/\partial\nu$ denotes the outward normal derivative).

We shall assume that

(H1) $f : G \subset \mathbb{R}^n \longrightarrow \mathbb{R}^n$ is a locally Lipschitz continuous function on an open subset G of $\mathbb{R}^n$.

(H2) Ω is a bounded open connected subset of $\mathbb{R}^m$, $m \in \mathbb{N} - \{0\}$ with a sufficiently smooth boundary $\partial\Omega$.

(H3) β_i , α_i are sufficiently smooth nonnegative functions on $\partial\Omega$ such that $\alpha_i + \beta_i > 0$ on $\partial\Omega$. $\beta_i = 0$ means homogeneous Dirichlet boundary conditions, while $\alpha_i = 0$ means homogeneous Neumann boundary conditions.

(H4) $D = diag(d_i)$, $d_i > 0$.

Denote by X the real Banach space $C(\overline{\Omega})$ of continuous vector valued functions $u : \overline{\Omega} \longrightarrow \mathbb{R}^n$, endowed with the norm

$$\|u\| = \sum_{i=1}^{n} \sup_{x \in \overline{\Omega}} |u_i(x)| \ .$$

The space X is a (partially) ordered Banach space with respect to the cone

$$X_+ := \{v \in X \mid 0 \leq v(x),\ x \in \overline{\Omega}\}$$

Note that the order $\leq$ induced on X by the cone X_+, is compatible with the order $\leq$ induced on $\mathbb{R}^n$ by the cone $\mathbb{K}$; i.e.

$$u \leq v \quad \text{in} \quad X \Longleftrightarrow u(x) \leq v(x) \quad \text{in } \mathbb{R}^n \ , \quad \text{for any } x \in \overline{\Omega} \ .$$

In an analogous way we may then define $u < v$ and $u \ll v$ in X, by the correspondence in $\mathbb{R}^n$.

As for the finite dimensional case, in the Banach space X we may define a "(local) semiflow" as a mapping

$$\varphi \ : \ S \subset \mathbb{R}_+ \times X \ \longrightarrow \ X$$

where

$$S = \left\{(t,u) \in \mathbb{R}_+ \times X \ \middle| \ t \in J(u) := \left[0, \tau^+(u)\right) \right\}$$

($0 < \tau^+(u) \leq +\infty$), having the following properties

(i) $\varphi(0;u) = u$, for any $u \in X$

(ii) $\varphi(t+s;u) = \varphi(t;\varphi(s;u))$, for any $u \in X$ and for all $s \in J(u)$, and $t \in J(\varphi(s;u))$

(iii) the mapping $t \longrightarrow \varphi(t;u)$ is continuous on $J(u)$, for any $u \in X$

(iv) the mapping $u \longrightarrow \varphi(t;u)$ is continuous in X, for any $t \in J(u)$.

Theorem B.17. [169] *Assume conditions (H1)-(H4) hold. Then problem (B.1), (B.1b), (B.1o) defines a (local) semiflow φ on the Banach space X. For any $u_0 \in X$, $\{\varphi(t;u_0),\ t \in J(u)\}$ provides the unique solution of the problem (B.1), (B.1b), (B.1o) (for homogeneous Dirichlet boundary conditions $X := \{u \in C(\overline{\Omega}) \mid u = 0 \text{ on } \partial\Omega\}$). This semiflow satisfies the following properties.*

(a) (Maximality) For any $u_0 \in X$, such that $\tau^+(u_0) < +\infty$, we have

$$\lim_{t \to \tau^+(u_0)} \|\varphi(t;u_0)\| = +\infty \ .$$

(b) *(Compactness) If $u_0 \in X$ is such that the positive orbit $\Gamma_+(u_0) := \{\varphi(t;u_0) \mid t \in \mathbb{R}_+\}$ is bounded in X, then $\Gamma_+(u_0)$ is relatively compact.*

A boundary operator B may be defined as

$$(Bu)_i := \beta_i \frac{\partial}{\partial \nu} u_i + \alpha_i u_i \;, \quad i = 1, \cdots, n \quad \text{on } \partial\Omega \;.$$

Global boundedness of solutions of (B.1), (B.1b) may follow e.g. from the existence of a bounded invariant region.

A positively invariant region for the local semiflow defined by (B.1), (B.1b) is a closed subset $\Sigma \subset \mathbb{R}^n$ such that for every initial state $u_0 \in X$, having $u_0(x) \in \Sigma$ for any $x \in \overline{\Omega}$, $\varphi(t;u_0)(x)$ remains in Σ for any $x \in \overline{\Omega}$ and $t \in J(u_0)$.

Actually when we deal with general third type boundary conditions we need to be more precise; an invariant region for the initial-boundary value problem (B1), (B1b) is a set $\Sigma \subset \mathbb{R}^n$ such that if $\{u(x;t)\;,\; x \in \overline{\Omega}\;, t \geq 0\}$ is a solution of (B1) with $u_0(x) \in \Sigma$, for $x \in \Omega$, and $Bu \in \alpha\Sigma$ on $\partial\Omega \times \mathbb{R}_+$ then $u(x;t) \in \Sigma$ for $x \in \overline{\Omega}$, $t \in \mathbb{R}_+$ [39, 204]; here $\alpha\Sigma := \{(\alpha_1 v_1, \cdots, \alpha_n v_n)^T \mid v \in \Sigma\}$.

Typically we consider regions Σ of the form

$$\Sigma := \bigcap_{j=1}^{p} \{z \in \mathbb{R}^n \mid G_j(z) \leq 0\}$$

where G_j are suitable smooth functions on G.

Theorem B.18. [204] *If, at each point $z \in \partial\Sigma$ we have*

(a) *$\nabla G_j(z)$ is a left eigenvector of D, $j = 1, \cdots, p$;*

(b) *$\nabla G_j(z) \cdot w = 0$ for any $w \in \mathbb{R}^n \implies w \cdot H(z)\, w \geq 0$ for any $w \in \mathbb{R}^n$, with H the Hessian matrix of G_j, $j = 1, \cdots, p$;*

(c) *$\nabla G_j(z) \cdot f(z) < 0$, $j = 1, \cdots, p$;*

then Σ is positively invariant for (B.1), (B.1b).
Conditions (a)-(c) are also necessary conditions for the invariance of Σ, with a weak inequality in (c).

Remark. Provided the vector field f points into Σ on $\partial\Sigma$ we have : (i) if D is a scalar matrix ($D = dI$, $d \in \mathbb{R}_+^*$), and Σ is convex then it is invariant; (ii) otherwise Σ is invariant if and only if it is a (possibly unbounded) parallelogram. The edges of this parallelogram are parallel to the coordinate axes if and only if D is a diagonal matrix. In this case then the invariance of the positive cone X_+ is allowed provided the vector field f points into $\mathbb{K}$ on $\partial\mathbb{K}$.

In the sequel we shall assume conditions on f that allow the global existence of the semiflow for any $u_0 \in X_+$. Further we shall assume that f is smooth enough so that the solutions u of system (B.1), (B.1b) are classical solutions, i.e.

$$u \in C^{2,1}\left(\Omega \times (0,+\infty), \mathbb{R}^n\right) \cap C^{1,0}\left(\overline{\Omega} \times (0,+\infty), \mathbb{R}^n\right)$$

and satisfies the system in a classical sense.

If we denote by $\{U(t)\,u_0\ ,\ t \in \mathbb{R}_+\}$ the unique solution of system (B.1), (B.1b), (B.1o) it can be shown [91] that the evolution operator $\{U(t)\,,\ t \in \mathbb{R}_+\}$ is a (nonlinear) C_o-semigroup on X (see Section B.1.4).

Note that whenever we do not have global (in time) solutions, properties (i)-(iii) of Section B.1.4 apply to the maximal interval of existence.

B.2.1. Semilinear quasimonotone parabolic autonomous systems

Here we extend comparison theorems to the parabolic case.

Theorem B.19. [39, 161, 86] *Under the assumptions of Theorem B.17, let further f in system (B.1) be quasimonotone nondecreasing (cooperative) in $\mathbb{K}$. Let $u(x;t) = (u_1(x;t),\cdots,u_n(x;t))^T$ and $v(x;t) = (v_1(x;t),\cdots,v_n(x;t))^T$ be classical solutions of the following two inequalities, respectively,*

$$(B.2) \qquad \frac{\partial}{\partial t}u \leq D\,\Delta u + f(u)$$

$$(B.3) \qquad \frac{\partial}{\partial t}v \geq D\,\Delta v + f(v)\ ,$$

in $\Omega \times (0,+\infty)$, with boundary conditions

$$(B.4) \qquad \beta_i\frac{\partial}{\partial \nu}u_i + \alpha_i u_i \leq \beta_i\frac{\partial}{\partial \nu}v_i + \alpha_i v_i\ , \quad i = 1,\cdots,n\ ,$$

on $\partial\Omega \times (0,+\infty)$, and initial conditions such that

$$(B.5) \qquad u_0 \leq v_0 \qquad in \ \overline{\Omega} \ .$$

Then

$$(B.6) \qquad u \leq v \qquad in \ \overline{\Omega} \times \mathbb{R}_+ \ .$$

The above theorem implies in particular the monotonicity of the evolution operator

$$(B.7) \qquad u_0\,,\ v_0 \in X_+\,,\ u_0 \leq v_0 \quad \Longrightarrow \quad U(t)\,u_0 \leq U(t)\,v_0 \ , \quad t \geq 0 \ .$$

If we further assume for $f \in C^1(\mathbb{K}, \mathbb{R}^n)$ the hypotheses (F1)-(F3) of Section A.4.2, then first of all we may actually claim that for any choice of $u_0 \in X_+$, a unique global solution exists for system (B1), (B.1b) subject to the initial condition u_0 at $t = 0$. Hence the family of evolution operators $\{U(t)\,,\ t \in \mathbb{R}_+\}$ define a (global) flow in X_+.

Moreover we may state the following

$$(B.8) \qquad U(t)\,0 = 0 \ , \qquad t \geq 0$$

$$(B.9) \qquad U(t)\,X_+ \subset X_+ \ , \qquad t \geq 0$$

Lemma B.20. [161] *Under the assumptions of Lemma A.33 the evolution operator $U(t)$ of system (B.1), (B.1b) is strongly positive for any $t > 0$; i.e.*

$$(B.10) \qquad U(t)\,(\dot{X}_+) \subset \overset{\circ}{X}_+ \ , \qquad t > 0$$

Lemma B.21. [161] *Under the assumptions of Lemma A.33*

(B.11) *for any* $u_0\,,\ v_0 \in X_+$

$$u_0 \leq v_0\,,\ u_0 \neq v_0 \quad \Longrightarrow \quad U(t)\,u_0 \ll U(t)\,v_0$$

Lemma B.22. [161] *Under the assumptions of Lemma A.34, the evolution operator $U(t)$ of system (B.1), (B.1b) is strongly concave, for any $t > 0$.*

(B.12) *for any* $u_0 \in \overset{\circ}{X}_+$, *and for any* $\sigma \in (0,1)$ *an* $\alpha = \alpha(u_0, \sigma) > 0$

exists such that

$$U(t)\,(\sigma u_0) \geq (1+\alpha)\,\sigma\, U(t)\, u_0\ , \quad t>0\ .$$

Thus Theorem A.36 applies also for the PDE case, excluding the existence of more than one nontrivial equilibrium for system (B.1), (B.1b) under the assumptions of Lemma B.22.

B.2.1.1. The linear case

Let B be a real $n \times n$ quasimonotone (cooperative) matrix, i.e. such that $b_{ij} \geq 0$ for $i \neq j\,,\ \ i, j = 1, \cdots, n$, and consider the linear quasimonotone parabolic system

$$(B.13) \qquad \frac{\partial}{\partial t} v = D\Delta v + Bv\ , \qquad \text{in } \Omega \times (0,+\infty)$$

subject to the boundary conditions (B.1b).

The evolution operator associated with system (B.13), (B.1b) will be denoted by $\{T(t)\,,\ t\in \mathbb{R}_+\}$; it generates a (global) semiflow in X_+. Moreover, since equation (B.13) is linear, $T(t)$ will be, for any $t \in \mathbb{R}_+$, a linear operator.

Theorem B.23. [1] *Consider the following eigenvalue problem for the Laplace operator Δ (actually any strongly uniformly elliptic operator)*

$$(B.14) \qquad \begin{cases} \Delta\phi + \lambda\phi = 0\ , & in\ \ \Omega \\ \beta\dfrac{\partial}{\partial \nu}\phi + \alpha\phi = 0\ , & on\ \partial\Omega \end{cases}$$

where $\alpha\,,\ \beta$ are sufficiently smooth functions in $\partial\Omega$, itself sufficiently smooth. The eigenvalue problem (B.14) admits a smallest eigenvalue λ_α which is real and nonnegative. A unique (normalized) eigenfunction ϕ_α is associated with λ_α and it can be chosen strictly positive ($\phi_\alpha \gg 0$ in Ω). If $\alpha \geq 0$ on $\partial\Omega$, is not identically zero, then $\lambda_\alpha > 0$; if $\alpha \equiv 0$ on $\partial\Omega$, then $\lambda_\alpha = 0$. If $\beta \neq 0\,,\ \phi_\alpha \gg 0$ in $\overline{\Omega}$.

Theorem B.24. (Separation of variables) [161] *Consider system (B.13) subject to the same boundary conditions on both components of v:*

$$(B.15) \qquad \beta\frac{\partial}{\partial \nu} v + \alpha v = 0 \qquad on\ \partial\Omega \times (0,+\infty)$$

where α, $\beta \geq 0$ are sufficiently smooth real functions defined in $\partial\Omega$. Suppose that the (B.13), (B.15) is subject to an initial condition of the form

$$(B.16) \qquad v(x;0) = v_0(x) = \phi_\alpha(x)\,\xi\ , \qquad \text{in } \overline{\Omega}$$

where ϕ_α is the unique eigenfunction associated with the first eigenvalue of problem (B.14), and $\xi \in \mathbb{K}$, then the solution of system (B.13), (B.15), (B.16) is given by

$$(B.17) \qquad v(x;t) = [T_\alpha(t)\,v_0]\,(x) = \phi_\alpha(x)\,w_\xi(t)$$

in $\overline{\Omega} \times [0,+\infty)$, where $w_\xi(t)$, $t \geq 0$ is the unique solution of the following ODE system

$$(B.18) \qquad \frac{d}{dt}w = (-\lambda_\alpha D + B)w\ , \qquad t > 0$$

subject to the initial condition

$$(B.19) \qquad w(0) = \xi\ .$$

Note that, in the above theorem, we have denoted by $\{T_\alpha(t)\,,\ t \in \mathbb{R}_+\}$ the evolution semigroup of linear operators associated with system (B.13), (B.15).

Lemma B.25. *Let $v(t) = T(t)\,v_0$, $t \in \mathbb{R}_+$ be the solution of the linear system (B.13), (B.1b) subject to the initial condition $v_0 \in X_+$. If B is quasimonotone (cooperative) irreducible then*

$$(B.20) \qquad v_0 \in X_+\ ,\ v_0 \neq 0 \implies v(t) = T(t)\,v_0 \gg 0\ , \qquad t > 0$$

i.e.

$$(B.20') \qquad T(t)\,\dot{X}_+ \subset \overset{\circ}{X}_+\ .$$

Theorem B.26. *Consider system (B.13), (B.15).*

(i) If, for any $\lambda \in \sigma(-\lambda_\alpha D+B)$, we have $\mathcal{R}e\ \lambda < 0$, then the trivial solution is GAS in X_+ for system (B.13), (B.15)

(ii) if $\mu = \max\{\mathcal{R}e\ \lambda \mid \lambda \in \sigma(-\lambda_\alpha D + B)\} > 0$ and B is irreducible, then the trivial solution is unstable. Moreover

$$\text{(B.21)} \qquad \textit{for any} \quad v_0 \in \dot{X}_+ \ : \ \liminf_{t\to+\infty} \|T_\alpha(t)\, v_0\|\, e^{-\mu t} > 0\ .$$

B.2.1.2. The nonlinear case

From now on we shall denote by

$$\alpha_m = \min_{x\in\partial\Omega}\ \min_{1\le i\le n}\ \{\alpha_i(x)\}$$

$$\beta_m = \max_{x\in\partial\Omega}\ \max_{1\le i\le n}\ \{\beta_i(x)\}$$

and by

$$\alpha_M = \max_{x\in\partial\Omega}\ \max_{1\le i\le n}\ \{\alpha_i(x)\}$$

$$\beta_M = \min_{x\in\partial\Omega}\ \min_{1\le i\le n}\ \{\beta_i(x)\}$$

and by $\lambda_m\,(\phi_m)$, and $\lambda_M\,(\phi_M)$ the corresponding eigenvalues (eigenvectors) of problem (B.14).

Theorem B.27. [161] *Let B be a quasimonotone (cooperative) matrix such that*

$$\text{(B.22)} \qquad \textit{for any} \quad \xi \in \mathbb{K} \ : \ f(\xi) \le B\xi\ .$$

If, for any $\lambda \in \sigma(-\lambda_m D + B)$, we have $\mathcal{R}e\ \lambda < 0$, then the trivial solution is GAS in X_+ for system (B.1), (B.1b).

Theorem B.28. [161] *Let B be a quasimonotone (cooperative) irreducible matrix such that a $\delta > 0$ exists for which*

$$\text{for any} \quad \xi \in \mathbb{K}, \ \|\xi\| < \delta \ : \ f(\xi) \geq B\xi \ . \tag{B.23}$$

If, $\mu = \max\{\mathcal{R}e\ \lambda \mid \lambda \in \sigma(-\lambda_M D + B)\} > 0$ then the trivial solution is unstable for system (B.1), (B.1b).

B.2.1.3. Lower and upper solutions. Existence of nontrivial equilibria

If we deal with quasimonotone reaction-diffusion systems of the form (B.1), subject to homogeneous Neumann boundary conditions ($\alpha_i = 0\,,\ \beta_i \neq 0\,,\ i = 1, \cdots, n$, in (B.1b)), space homogeneous equilibria are possible, which are critical points of the vector function f.

Under this assumption it is not difficult to extend Theorem A.37 (Nested Invariant Rectangles) and Theorem A.38 (Nested Contracting Rectangles) to the PDE case [39].

When we have general boundary conditions (B.1b), space homogeneous equilibria are not any more allowed, and we are obliged to extend the definition of lower and upper solution to the PDE case, explicitly.

An equilibrium solution of system (B.1), (B.1b) is a classical solution $\phi \in C^2(\Omega) \cap C^1(\overline{\Omega})$ of system

$$D\,\Delta\phi + f(\phi) = 0\ , \qquad \text{in } \Omega \tag{B.24}$$

subject to the boundary condition

$$\beta_i \frac{\partial}{\partial \nu} u_i + \alpha_i u = 0\ , \qquad i = 1, \cdots, n\ , \qquad \text{on } \partial\Omega\ . \tag{B.1b}$$

A "lower solution" of system (B.24), (B.1b) is a classical solution $\underline{\phi} \in C^2(\Omega) \cap C^1(\overline{\Omega})$ of the following inequalities

$$D\,\Delta\underline{\phi} + f(\underline{\phi}) \geq 0\ , \qquad \text{in } \Omega \tag{B.25}$$

$$(B.25b) \qquad \beta_i \frac{\partial}{\partial \nu} \underline{\phi}_i + \alpha_i \underline{\phi}_i \leq 0 \ , \qquad i = 1, \cdots, n \ , \quad \text{on } \partial\Omega \ .$$

The notion of "upper solution" is obtained by reversing the inequalities in (B.25) and (B.25b).

Lemma B.29. [161] *Let* $\underline{\phi} \geq 0$ *be a lower solution of system (B.25), (B.25b). Then* $\{U(t)\,\underline{\phi},\ t \in \mathbb{R}_+\}$ *is monotone nondecreasing in* t. *Further, if we set*

$$(B.26) \qquad \begin{aligned} \mathcal{N}_+(\underline{\phi}) := \{\psi \in C^2(\Omega) \cap C^1(\overline{\Omega}) \mid \psi \ \textit{is an equilibrium} \\ \textit{for } (\text{B.1}), (\text{B.1b}), \ \textit{and } \underline{\phi} \leq \psi\} \end{aligned}$$

and $\mathcal{N}_+(\underline{\phi}) \neq \emptyset$, *then a* $\phi_- = \min \mathcal{N}_+(\underline{\phi})$ *exists such that*

$$lim_{t \to +\infty}\, dist\left(U(t)\,\underline{\phi}\ ,\ \phi_-\right) = 0$$

Lemma B.30. [161] *Let* $\overline{\phi} \leq 0$ *be an upper solution of system (B.25), (B.25b). Then* $\left\{U(t)\,\overline{\phi},\ t \in \mathbb{R}_+\right\}$ *is monotone nonincreasing in* t. *Moreover, if we set*

$$(B.27) \qquad \begin{aligned} \mathcal{N}_-(\overline{\phi}) := \{\psi \in C^2(\Omega) \cap C^1(\overline{\Omega}) \mid \psi \ \textit{is an equilibrium} \\ \textit{for } (\text{B.1}), (\text{B.1b}), \ \textit{and } \psi \leq \overline{\phi}\}\ , \end{aligned}$$

then a $\phi_+ = \max \mathcal{N}_-(\overline{\phi})$ *exists such that*

$$\lim_{t \to +\infty} dist\left(U(t)\,\overline{\phi}\ ,\ \phi_+\right) = 0\ .$$

Lemma B.31. [161] *Let* $(\chi_k)_{k \in \mathbb{N}}$ *be a monotone increasing sequence of equilibria of system (B.1), (B.1b) such that for any* $k \in \mathbb{N}$, $\chi_k \gg 0$ *in* $\overline{\Omega}$ *(in* Ω *for homogeneous Dirichlet boundary conditions). Then a* $\chi \gg 0$ *in* $\overline{\Omega}$ *(in* Ω *for homogeneous Dirichlet boundary conditions) exists such that* $\lim_{k \to +\infty} \chi_k = \chi$, *in* X. χ *is itself an equilibrium of system (B.1), (B.1b).*

Theorem B.32. [137, 139, 161] *Let* f *in system (B.1) satisfy the following assumptions*

(i) a quasimonotone (cooperative), irreducible matrix B *exists for which a* $\delta > 0$ *exists such that*

$$\textit{for any} \quad \xi \in \mathbb{K}, \ \|\xi\| < \delta \ : \ f(\xi) \geq B\xi$$

and

$$\mu = \max\{\mathcal{R}e\ \lambda \mid \lambda \in \sigma(-\lambda_M D + B)\} > 0$$

(ii) *a quasimonotone (cooperative) matrix C exists for which a $\delta > 0$ exists such that*

$$\textit{for any} \quad \xi \in \mathbb{K}, \ \|\xi\| \geq \delta \ : \ f(\xi) \leq C\xi$$

and

$$\textit{for any} \quad \lambda \in \sigma(C) \ : \ \mathcal{R}e\ \lambda < 0$$

(iii) *(F4) and (F5) of Section A.4.2 .*

Then system (B.1), (B.1b) admits a unique nontrivial equilibrium solution $\phi \gg 0$ in $\overline{\Omega}$ (in Ω for homogeneous Dirichlet boundary conditions) which is GAS in $\dot{X}_+$.

B.2.2. Lyapunov methods for PDE 's, LaSalle Invariance Principle in Banach space

Suppose we are given a semidynamical system on a closed subset D of a Banach space X.

As we have seen in the previous sections it may be defined by a C_o-semigroup of nonlinear operators $\{U(t)\,;\ t \in \mathbb{R}_+\}$ acting on D (see Section B.1.4).

For any $u \in D$, we may define the positive orbit starting from u at $t = 0$, as usual

$$\Gamma_+(u) := \{U(t)\, u \in D \mid t \in \mathbb{R}_+\} \ .$$

We say that $\phi \in D$ is an equilibrium point for $\{U(t)\,,\ t \in \mathbb{R}_+\}$ if $\Gamma_+(\phi) = \phi$.

Stability concepts can be rephrased in a Banach space, with respect to its norm.

A "Lyapunov functional" for the dynamical system $\{U(t)\,;\ t \in \mathbb{R}_+\}$ on $D \subset X$ is a continuous real valued function $\mathcal{V} : D \subset X \longrightarrow \mathbb{R}$ such that

$$\dot{\mathcal{V}}(u) := \limsup_{t \to 0+} \frac{1}{t} \{\mathcal{V}(U(t)\, u) - \mathcal{V}(u)\} \leq 0$$

for all $u \in D$.

Theorem B.33. [108] *Let $\{U(t)\,,\ t \in \mathbb{R}_+\}$ be a dynamical system on D, and let 0 be an equilibrium point in D. Suppose $\mathcal{V}$ is a Lyapunov function on D which satisfies*

(i) $\mathcal{V}(0) = 0$

(ii) $\mathcal{V}(u) \geq c(\|u\|)$, $u \in D$ *where c is a continuous strictly increasing function such that $c(0) = 0$ and $c(r) > 0$ for $r > 0$.*

Then 0 is stable. Suppose in addition that

(iii) $\dot{\mathcal{V}}(u) \leq -c_1(\|u\|)$, $u \in D$ *where c_1 is also continuous increasing and positive with $c_1(0) = 0$.*

Then 0 is uniformly asymptotically stable.

A set $\Sigma \subset D$ is "positively invariant" for the dynamical system $\{U(t)\,,\ t \in \mathbb{R}_+\}$ if $U(t)\,\Sigma \subset \Sigma$, for any $t \in \mathbb{R}_+$.

If $u_0 \in D$ and $\Gamma_+(u_0)$ is its positive orbit, then the ω-limit set of u_0, or of $\Gamma_+(u_0)$, is

$$\omega(u_0) = \omega(\Gamma_+(u_0)) := \{u \in D \mid \text{a sequence } t_n \in \mathbb{R}_+ \text{ exists such that } t_n \longrightarrow \infty \text{, and } U(t_n)u_0 \longrightarrow u \text{, for } n \longrightarrow \infty\} .$$

Theorem B.34. [108, 215] *Suppose* $u_0 \in D$ *is such that its orbit* $\Gamma_+(u_0)$ *is precompact (lies in a compact set of* D*); then* $\omega(u_0)$ *is nonempty, compact, invariant and connected. Moreover*

$$\lim_{t \to +\infty} dist(U(t)\, u_0 \, , \, \omega(u_0)) = 0 \, .$$

Theorem B.35. (LaSalle Invariance Principle) [108, 215] *Let* $\mathcal{V}$ *be a Lyapunov functional on* D *(so that* $\dot{\mathcal{V}}(u) \leq 0$ *on* D*). Define*

$$E := \left\{ u \in D \;\middle|\; \dot{\mathcal{V}}(u) = 0 \right\}$$

and let M *be the largest (positively) invariant subset of* E. *If for* $u_0 \in D$ *the orbit* $\Gamma_+(u_0)$ *is precompact (lies in a compact set of* D*), then*

$$\lim_{t \to +\infty} dist(U(t)\, u_0 \, , \, M) = 0 \, .$$

Remark. For dynamical systems generated by evolution equations

$$\frac{d}{dt} u = Au + f(u)$$

where A is a strongly elliptic operator, bounded orbits are generally precompact [108, 219], and boundedness of orbits frequently follows from the existence of a Lyapunov functional such that $\{u \in D \mid \mathcal{V}(u) < k\}$ is a bounded set for a suitable choice of $k > 0$.

References

[1] Amann, H. : Fixed point equations and nonlinear eigenvalue problems in ordered Banach spaces. SIAM Review **18** (1976),620-709.

[2] Amann, H. : Periodic solutions of semilinear parabolic equations. In : Cesari,L.,Kannan.R.,Weinberger,H.(eds.) Nonlinear Analysis. A collection of Papers in Honor of Erich Rothe. Academic Press, New York, 1978, pp. 1–29.

[3] Amann, H. : Ordinary Differential Equations. de Gruyter, Berlin, 1990.

[4] Anderson, R.M. : The dynamics and control of direct life cycle helminthic parasites. In : Barigozzi, C. (ed.) Vito Volterra Symposium on Mathematical Models in Biology. (Lecture Notes in Biomathematics, vol. 39.) Springer-Verlag, Berlin, Heidelberg, 1980.

[5] Anderson, R.M. : Directly transmitted viral and bacterial infections of man. In : Anderson, R.M. (ed.) Population Dynamics of Infectious Diseases. Theory and Applications. Chapman and Hall, New York, 1982, pp. 1–37.

[6] Anderson, R.M., May, R.M. : Population biology of infectious diseases. Part I, Nature **280** (1979), 361–367.

[7] Anderson, R.M., May, R.M. : Population biology of infectious diseases. Part II, Nature **280** (1979), 455–461.

[8] Anderson, R.M., May, R.M. : The population dynamics of microparasites and their invertebrate hosts. Trans. R. Philos. Soc. B, **291** (1981), 451–524.

[9] Anderson, R.M., May, R.M. : Infectious Diseases of Humans. Dynamics and Control. Oxford University Press, Oxford, 1991.

[10] Anderson, R.M., May, R.M., Medley, G.F., Johnson, A. : A preliminary study of the transmission dynamics of the human immunodeficiency virus (HIV), the causative agent of AIDS. IMA J. Math. Med. Biol. **3** (1986), 229–263.

[11] Anderson, D., Watson, R. : On the spread of a disease with gamma distributed latent and infectious periods. Biometrika **67** (1980), 191–198.

[12] Aris, R. : The Mathematical Theory of Diffusion and Reaction in Permeable Catalysts. Vols. I and II, Oxford Univ. Press, Oxford, 1975.

[13] Arnautu, V., Barbu, V., Capasso, V. : Controlling the spread of epidemics. Appl. Math. Optimiz. **20** (1989), 297–317.

[14] Aron, J.L., May, R.M. : The population dynamics of malaria. In : Anderson, R.M. (ed.) Population Dynamics of Infectious Diseases. Theory and Applications. Chapman and Hall, New York, 1982.

[15] Aronson, D.G. : The asymptotic speed of propagation of a simple epidemic. In : Fitzgibbon,W.E. and Walker, A.F. (eds.) Nonlinear Diffusion. Pitman, London, 1977.

[16] Aronson, D.G. : The role of diffusion in mathematical population biology : Skellam revisited. In : Capasso, V., Grosso, E. and Paveri-Fontana, S.L. (eds.) Mathematics in Biology and Medicine. (Lecture Notes in Biomathematics, vol. 57.) Springer-Verlag, Berlin, Heidelberg, 1985.

[17] Aronson, D.G., Weinberger, H.F. : Nonlinear diffusions in population genetics, combustion and nerve propagation. In : Goldstein, J. (ed.) Partial Differential

Equations and Related Topics. (Lecture Notes in Mathematics, vol. 446.) Springer-Verlag, Berlin, Heidelberg, 1975.

[18] Aronsson, G., Mellander, I. : A deterministic model in biomathematics: asymptotic behavior and threshold conditions. Math. Biosci. **49** (1980), 207–222.

[19] Bailey, N.T.J. : The Mathematical Theory of Infectious Diseases. Griffin, London, 1975.

[20] Bailey, N.T.J. : The Biomathematics of Malaria. Griffin, London, 1975.

[21] Banks, H.T., Mahaffy, J.M. : Global asymptotic stability synthesis and repression. Quart. Appl. Math. **36** (1978), 209–221.

[22] Barbour, A.D. : Macdonald's model and the transmission of bilharzia. Trans. Roy. Soc. Trop. Med. Hyg. **72** (1978), 6–15.

[23] Barbour, A.D. : Schistosomiasis. In : Anderson, R.M. (ed.) Population Dynamics of Infectious Diseases. Chapman and Hall, London, 1982.

[24] Basile, N., Mininni, M. : An extension of the maximum principle for a class of optimal control problems in infinite dimensional spaces. SIAM J. Control and Optimization **28** (1990),1113-1135.

[25] Basile, N., Mininni, M. : A vector valued optimization approach to the study of a class of epidemics. J. Math. Anal. Appl. **155** (1990),485-498.

[26] Belleni-Morante, A. : Applied Semigroups and Evolution Equations. Oxford University Press, Oxford, 1979.

[27] Bellman, R. : Introduction to Matrix Analysis. McGraw-Hill, New York, 1960.

[28] Beretta, E., Capasso, V. : On the general structure of epidemic systems. Global asymptotic stability. Comp. and Maths. with Appls. **12A** (1986), 677–694.

[29] Beretta, E., Capasso, V. : Global stability results for a multigroup SIR epidemic model. In : Gross L.J., Hallam T.G. and Levin S.A. (eds.) Mathematical Ecology. World Scientific, Singapore, 1988.

[30] Beretta, E., Capasso, V., Rinaldi, F. : Global stability results for a generalized Lotka-Volterra system with distributed delays. Application to predator-prey and to epidemic systems. J. Math. Biol. **26** (1988), 661–688.

[31] Berman, A., Plemmons, R.J. : Nonnegative Matrices in the Mathematical Sciences. Academic Press, New York, 1979.

[32] Bernoulli, D. : Essai d' une nouvelle analyse de la mortalité causée par la petite vérole et des avantages de l'inoculation pour la prévenir. Mem. Math. Phys. Acad. Roy. Sci., (1760), 1–45.

[33] Blat, J., Brown, K.J. : A reaction-diffusion system modelling the spread of bacterial infections. Math. Meth. Appl. Sci. **8** (1986), 234–246.

[34] Blythe, S.P., Anderson, R.M. : Distributed incubation and infectious periods in models of the transmission dynamics of the human immuno-deficiency virus (HIV). IMA J. Math. Med. Biol. **5** (1988), 1–19.

[35] Brauer, F. : Epidemic models in populations of varying population size. In: Castillo-Chavez, C., Levin, S.A. and Shoemaker C.A. (eds.) Mathematical Approaches to Problems in Resource Management and Epidemiology. (Lec-

ture Notes in Biomathematics, vol. 81.) Springer-Verlag, Berlin, Heidelberg, 1989, pp. 109–123.

[36] Brauer, F. : Some infectious disease models with population dynamics and general contact rates. Differential and Integral Equations, **3** (1990), 827–836.

[37] Brauer, F. : Models for the spread of universally fatal diseases. J. Math. Biol. **28** (1990), 451–462.

[38] Brauer, F., Nohel, J.A. : Qualitative Theory of Ordinary Differential Equations. Benjamin, New York, 1969.

[39] Britton, N.F., Reaction-Diffusion Equations and Their Applications to Biology. Academic Press, London, 1986.

[40] Busenberg, S., Cooke, K. : The Dynamics of Vertically Transmitted Diseases. To appear.

[41] Busenberg, S., Cooke, K., Iannelli, M. : Endemic thresholds and stability in a class of age-structured epidemics. SIAM J. Appl. Math. **48** (1988), 1379–1395.

[42] Busenberg, S.N., Iannelli, M., Thieme, H.R. : Global behavior of an age structured SIS epidemic model. Preprint U.T.M. **282**, Trento, 1989.

[43] Busenberg, S.N., Iannelli, M., Thieme, H.R. : Global behavior of an age structured SIS epidemic model. The case of a vertically transmitted disease. Preprint U.T.M. **308**, Trento, 1990.

[44] Busenberg, S., van den Driessche, P. : Analysis of a disease transmission model in a population with varying size. J. Math. Biol. **28** (1990), 257–270.

[45] Capasso, V. : Global solution for a diffusive nonlinear deterministic epidemic model. SIAM J. Appl. Math. **35** (1978), 274–284.

[46] Capasso, V. : Asymptotic stability for an integro-differential reaction- diffusion system. J. Math. Anal. Appl. **103** (1984), 575–588.

[47] Capasso, V. : A counting process approach for age-dependent epidemic systems. In : Gabriel, J.P., Lefevre, C. and Picard, P. (eds.) Stochastic Processes in Epidemic Theory. (Lecture Notes in Biomathematics, vol. 86.) Springer-Verlag, Berlin, Heidelberg, 1990, pp. 118–128.

[48] Capasso, V. : Mathematical modelling of transmission mechanisms of infectious diseases. An overview. In: Capasso, V. and Demongeot, J. (eds.) Proc. 1st European Conference on Math. Appl. Biology and Medicine , 1992.To appear.

[49] Capasso, V., Di Liddo, A. : Global attractivity for reaction-diffusion systems. The case of nondiagonal diffusion matrices. J. Math. Anal. Appl. **177** (1993),510-529.

[50] Capasso, V., Di Liddo, A. : Asymptotic behaviour of reaction-diffusion systems in population and epidemic models. The role of cross diffusion. J. Math. Biol. **32** (1994),453-463.

[51] Capasso, V., Doyle, M. : Global stability of endemic solutions for a multigroup SIR epidemic model. SASIAM Technical Report, Bari (I), 1990.

[52] Capasso, V., Forte, B. : Model building as an inverse problem in Biomathematics. In : Levin, S.A. (ed.) Frontiers in Mathematical Biology. (Lecture Notes in Biomathematics, vol. 100.) Springer-Verlag, Berlin, Heidelberg, 1993, pp. 600–608.

[53] Capasso, V., Fortunato, D. : Stability results and their applications to some reaction-diffusion problems. SIAM J. Appl. Math. **39** (1979), 37–47.

[54] Capasso, V., Fortunato, D. : Asymptotic behavior for a class of non autonomous semilinear evolution systems and application to a deterministic epidemic model. Nonlinear Anal., T.M.A. **4** (1979), 901–908.

[55] Capasso, V., Kunisch, K. : A reaction diffusion system arising in modelling man-environment diseases. Quart. Appl. Math. **46** (1988), 431–450.

[56] Capasso, V., Maddalena, L. : Asymptotic behavior for a system of nonlinear diffusion equations diseases. Rend. Accad. Sc. Fis. e Mat. Napoli **48** (1981), 475–495.

[57] Capasso, V., Maddalena, L. : Convergence to equilibrium states for a reaction-diffusion system modelling the spatial spread of a class of bacterial and viral diseases. J. Math. Biol. **13** (1981), 173–184.

[58] Capasso, V., Maddalena, L. : Saddle point behavior for a reaction-diffusion system. Application to a class of epidemic models. Math. Comp. Simulation **24** (1982), 540–547.

[59] Capasso, V., Maddalena, L. : Periodic solutions for a reaction-diffusion system modelling the spread of a class of epidemics. SIAM J. Appl. Math. **43** (1983), 417–427.

[60] Capasso, V., Paveri-Fontana, S.L. : A mathematical model for the 1973 cholera epidemic in the European Mediterranean regions. Rev. Epidem. et Sante' Publ. **27** (1979), 121–132. Errata, ibid. **28** (1980), 330.

[61] Capasso, V., Serio, G. : A generalization of the Kermack-McKendrick deterministic epidemic model. Math. Biosci. **42** (1978), 41–61.

[62] Capasso, V., Thieme, H. : A threshold theorem for a reaction-diffusion epidemic system. In : Aftabizadeh, R. (ed.) Differential Equations and Applications. Ohio Univ. Press, 1988.

[63] Capasso, V., Thieme, H. : In preparation.

[64] Casten, R.G., Holland, C.J. : Stability properties of solutions to systems of reaction-diffusion equations. SIAM J. Appl. Math. **33** (1977), 353–364.

[65] Castillo-Chavez, C., Cooke, K.L., Huang, W., Levin, S.A. : On the role of long incubation periods in the dynamics of acquired immunodeficiency syndrome (AIDS). Part 1 : Single population models. J. Math. Biol. **27** (1989), 373–398.

[66] Castillo-Chavez, C., Cooke, K.L., Huang, W., Levin, S.A. : On the role of long incubation periods in the dynamics of acquired immunodeficiency syndrome (AIDS). Part 2 : Multiple group models. In : Castillo-Chavez, C. (ed.) Mathematical and Statistical Approaches to AIDS Epidemiology. (Lecture Notes in Biomathematics, vol. 83.) Springer-Verlag, Heidelberg, 1989.

[67] Castillo-Chavez, C., Cooke, K.L., Huang, W., Levin, S.A. : Results on the dynamics for models for the sexual transmission of the human immunodeficiency virus. Appl. Math. Lett. **2** (1989), 327–331.

[68] Chafee, N. : Behavior of solutions leaving the neighborhood of a saddle point for a nonlinear evolution equation. J. Math. Anal. Appl. **58** (1977), 312–325.

[69] Cliff, A.D., Haggett, P., Ord, J.K., Versey, G.R. : Spatial Diffusion. An Historical Geography of Epidemics in an Island Community. Cambridge University Press, Cambridge, 1981.

[70] Coale, A.J. : The Growth and Structure of Human Populations. A Mathematical Investigation. Princeton University Press, Princeton, N.J., 1972.

[71] Cooke, K.L., Yorke, J.A. : Some equations modelling growth processes and gonorrhea epidemics. Math. Biosci. **16** (1973), 75–101.

[72] Coppel, W.A. : Stability and Asymptotic Behavior of Differential Equations. D.C. Heath, Boston, 1965.

[73] Cross, G.W. : Three types of matrix stability. Linear Algebra Appl. **20** (1978), 253–263.

[74] Cunningham, J. : A deterministic model for measles : Z. Naturforsch. **34c** (1979), 647–648.

[75] Cushing, J.M. : Integrodifferential Equations and Delay Models in Population Dynamics. Lecture Notes in Biomathematics, vol. 20, Springer-Verlag, Berlin, Heidelberg, 1977.

[76] Cvjetanovic, B., Grab, B., Uemura, K. : Epidemiological model of typhoid fever and its use in the planning and evaluation of antityphoid immunization and sanitation programmes. Bull. W.H.O. **45** (1975), 53–75.

[77] De Angelis, D.L., Post, W.M., Travis, C.C. : Positive Feedback in Natural Systems, Springer-Verlag, Berlin, Heidelberg, 1986.

[78] Diekmann, O. : Thresholds and travelling waves for the geographical spread of infection. J. Math. Biol. **6** (1978), 109–130.

[79] Diekmann, O., Hesterbeek, H., Metz, J.A.J. : On the definition and the computation of the basic reproduction ratio R_0 in models for infectious diseases in heterogeneous populations. J. Math. Biol. **28** (1990), 365–382.

[80] Dietz, K. : Transmission and control of arbovirus diseases. In : Ludwig, D. and Cooke, K.L. (eds.) Epidemiology. SIAM, Philadelphia, 1975.

[81] Dietz, K. : The incidence of infectious diseases under the influence of seasonal fluctuations. Lecture Notes in Biomathematics, vol. 11, Springer-Verlag, Berlin, Heidelberg, 1976, pp. 1–15.

[82] Dietz, K., Hadeler, K.P. : Epidemiological models for sexually transmitted diseases. J. Math. Biol. **26** (1988), 1–25.

[83] Dietz, K., Schenzle, D. : Mathematical models for infectious disease statistics. In : Atkinson, A.C. and Fienberg, S.E. (eds.) A Celebration of Statistics. Springer-Verlag, New York, 1985, pp. 167–204.

[84] Dietz, K., Schenzle, D. : Proportionate mixing models for age-dependent infection transmission. J. Math. Biol. **22** (1985), 117–120.

[85] Driver, R.D. : Ordinary and Delay Differential Equations. Springer-Verlag, New York, 1977.

[86] Fife, P.C. : Mathematical Aspects of Reacting and Diffusing Systems. Lecture Notes in Biomathematics, vol. 28, Springer-Verlag, Berlin, Heidelberg, 1979.

[87] Fife, P.C., Mc Leod J.B. : The approach of solutions of nonlinear diffusion equations to travelling front solutions. Arch. Rat. Mech. Anal. **65** (1977), 335–361.

[88] Francis, D.P., Feorino, P.M., Broderson, J.R., McClure, H.M., Getchell, J.P., McGrath, C.R., Swenson, B., McDougal, J.S., Palmer, E.L., Harrison, A.K., Barre-Sinoussi, F., Chermann, J.C., Montagnier, L., Curran, J.W., Cabradilla, C.D., Kalyanaraman, V.S. : Infection of chimpanzees with lymphadenopathy-associated virus. Lancet **2** (1984), 1276–1277.

[89] Frauenthal, J.C. : Mathematical Modelling in Epidemiology. Springer-Verlag, Berlin, Heidelberg, 1980.

[90] Freedman, H.I. : Deterministic Mathematical Models in Population Ecology. Marcel Dekker, New York, 1980.

[91] Friedman, A. : Partial Differential Equations. Holt, Rinehart and Winston, New York, 1969.

[92] Gabriel, J.P., Hanisch, H., Hirsch, W.M. : Dynamic equilibria of helminthic infections. In : Chapman, D.G. and Gallucci, V.F. (eds.) Quantitative Population Dynamics. Int. Co-op. Pub. House, Fairland, Maryland, 1981, pp. 83–104.

[93] Gantmacher, F.R. : Applications of the Theory of Matrices. Interscience, New York, 1959.

[94] Goh, B.S. : Global stability in two species interactions. J. Math. Biol. **3** (1976), 313–318.

[95] Goh, B.S. : Global stability in many species systems. Am. Nat. **111** (1977), 135–143.

[96] Goh, B.S. : Global stability in a class of predator-prey models, Bull. Math. Biol. **40** (1978), 525–533.

[97] Grabiner, D. : Mathematical models for vertically transmitted diseases. In : Busenberg, S. and Cooke, K.L. (eds.) The Dynamics of Vertically Transmitted Diseases. Springer-Verlag, Heidelberg,1992.

[98] Greenhalgh, D. : Analytical results on the stability of age-structured recurrent epidemic models. IMA J. Math. Appl. Med. Biol. **4** (1987), 109–144.

[99] Greenhalgh, D. : Threshold and stability results for an epidemic model with an age-structured meeting rate. IMA J. Math. Appl. Med. Biol. **5** (1988), 81–100.

[100] Griffith, J.S. : Mathematics of cellular control processes, II. Positive feedback to one gene. J. Theoret. Biol. **20** (1968), 209–216.

[101] Gripenberg, G. : On a nonlinear integral equation modelling an epidemic in an age-structured population. J. reine angew. Math. **341** (1983), 54–67.

[102] Gupta, N.K., Rink, R.E. : Optimal control of epidemics. Math. Biosci. **18** (1973), 383–396.

[103] Hadeler, K.P. : Diffusion Equations in Biology. In: Iannelli, M. (ed.) Mathematics of Biology. CIME, Liguori Editore, Napoli, 1979.

[104] Hale, J.K. : Ordinary Differential Equations. Wiley-Interscience, New York, 1969.

[105] Hamer, W.H. : Epidemic diseases in England. Lancet **1** (1906), 733-739.

[106] Hassell, M.P., May, R.M. : The population biology of host-parasite and host-parasitoid associations. In : Roughgarden, J., May, R.M. and Levin S.A. (eds.) Perspectives in Ecological Theory. Princeton University Press, Princeton, 1989, pp. 319–347.

[107] Hastings, A. : Global stability in Lotka-Volterra systems with diffusion. J. Math. Biol. **6** (1978), 163–168.

[108] Henry, D. : Geometric Theory of Semilinear Parabolic Equations. Lecture Notes in Mathematics, vol. 840. Springer-Verlag, Berlin, Heidelberg, 1981.

[109] Hernandez, J. : Branches of positive solutions for a reaction-diffusion system modelling the spread of a bacterial infection. IRMA-CNR Report, Bari, 1989.

[110] Hethcote, H.W. : Qualitative analyses of communicable disease models. Math. Biosci. **28** (1976), 335–356.

[111] Hethcote, H.W., Levin, S.A. : Periodicity in epidemiological models. In: Gross, L., Hallam, T.G., and Levin, S.A. (eds.). Springer-Verlag, Heidelberg,1989, pp.193-211.

[112] Hethcote, H.W., Lewis, M.A., van der Driessche, P. : An epidemiological model with a delay and a nonlinear incidence rate. J. Math. Biol. **27** (1989), 49–64.

[113] Hethcote, H.W., Stech, H.W., van den Driessche, P. : Periodicity and stability in epidemic models : a survey. In : Busenberg S. and Cooke, K.L. (eds.) Differential Equations and Applications in Ecology, Epidemics and Population Problems. Academic Press, New York, 1981, pp. 65–82.

[114] Hethcote, H.W., Thieme, H.R. : Stability of the endemic equilibrium in epidemic models with subpopulations. Math. Biosci. **75** (1985), 205–227.

[115] Hethcote, H.W., van Ark, J.W. : Epidemiological models for heterogeneous populations : proportionate mixing, parameter estimation and immunization programs. Math. Biosci. **84** (1987), 85–118.

[116] Hethcote, H.W., van den Driessche, P. : Some epidemiological models with nonlinear incidence. J. Math. Biol. **29** (1991), 271–287.

[117] Hethcote, H.W., Waltman, P. : Optimal vaccination schedules in a deterministic epidemic model. Math. Biosci. **18** (1973), 365–381.

[118] Hethcote, H.W., Yorke, J.A. : Gonorrhea Transmission Dynamics and Control. Lecture Notes in Biomathematics vol. 56, Springer-Verlag, Berlin, Heidelberg, 1984.

[119] Hirsch, W.M. : The dynamical systems approach to differential equations. Bull. Am. Math. Soc. **11** (1984), 1–64.

[120] Holling, G.S. : Some characteristics of simple types of predation and parasitism. Can. Ent. **91** (1959), 385–398.

[121] Hoppensteadt, F. : Mathematical Theories of Populations : Demographics, Genetics and Epidemics. SIAM, Philadelphia, 1975.

[122] Huang, W. : Studies in Differential Equations and Applications. Ph.D. Dissertation, The Claremont Graduate School, Claremont (CA), 1990.

[123] Inaba, H. : Threshold and stability results for an age-structured epidemic model. J. Math. Biology **28** (1990), 411–434.

[124] Jacquez, J.A., Simon, C.P., Koopman, J.S. : The reproduction number in deterministic models of contagious diseases. Comments in Theoretical Biology **2** (1991), 159–209.

[125] Jacquez, J.A., Simon, C.P., Koopman, J.S., Sattenspiel, L., Parry, T. : Modeling and analyzing HIV transmission : the effect of contact patterns. Math. Biosci. **92** (1988), 119–199.

[126] Jordan, P., Webbe, G. : Human Schistosomiasis. Heineman, London, 1969.

[127] Kamke, E. : Zur Theorie der Systeme gewöhnlicher Differentialgleichungen, II, Acta Math. **58** (1932), 57–85.

[128] Kamke, E. : Differentialgleichungen: Lösungsmethoden und Lösungen I. Teubner, Stuttgart,1977.

[129] Kato, T. : Perturbation Theory for Linear Operators. Springer-Verlag, New York, 1966.

[130] Kendall, D.G. in discussion on "Bartlett, M.S., Measles periodicity and community size", J. Roy. Stat. Soc. Ser. A, **120** (1957), 48–70.

[131] Kendall, D. G. : Mathematical models of the spread of infection. In : Mathematics and Computer Science in Biology and Medicine. H.M.S.O., London, 1965, pp. 213–225.

[132] Kermack, W.O., McKendrick, A.G. : Contributions to the mathematical theory of epidemics.
- Part I, Proc. Roy. Soc., A, **115** (1927), 700–721,
- Part II, Proc. Roy. Soc., A, **138** (1932), 55–83,
- Part III, Proc. Roy. Soc., A, **141** (1933), 94–122,
- Part IV, J. Hyg. Camb. **37** (1937), 172–187,
- Part V, J. Hyg. Camb. **39** (1939), 271–288.

[133] Kolesov, Ju. S. : Certain tests for the existence of stable periodic solutions of quasilinear parabolic equations. Soviet Math. Doklady **5** (1964), 1118–1120.

[134] Kolesov, Ju. S. : Periodic solutions of quasilinear parabolic equations of second order. Trans. Moscow Math. Soc. **21** (1970), 114–146.

[135] Kolesov, Ju. S., Krasnoselskii, M.A. : Lyapunov stability and equations with concave operators. Soviet Math. Doklady **3** (1962), 1192–1196.

[136] Krasnoselskii, M.A. : Stability of periodic solutions emerging from an equilibrium state. Soviet Math. Doklady **4** (1963), 679–682.

[137] Krasnoselskii, M.A. : Positive Solutions of Operator Equations. Noordhoff, Groningen, 1964.

[138] Krasnoselskii, M.A. : The theory of periodic solutions of nonautonomous differential equations. Russian Math. Surveys **21** (1966), 53–74.

[139] Krasnoselskii, M.A. : Translation along Trajectories of Differential Equations. Transl. of Math. Monographs vol. 19, Am. Math. Soc., Providence, R.I., 1968.

[140] Kreyszig, E. : Introductory Functional Analysis with Applications. Wiley, New York, 1978.

[141] Kunisch, K., Schelch, H. : Parameter estimation in a special reaction- diffusion system modelling man-environment diseases. J. Math. Biol. **27** (1989), 633-665.

[142] Lajmanovich, A., Yorke, J.A. : A deterministic model for gonorrhea in a nonhomogeneous population. Math. Biosci. **28** (1976), 221–236.

[143] Lakshmikantham, V. : Comparison results for reaction-diffusion equations in a Banach space. Conf. Sem. Mat. Univ. Bari, **158–162** (1979), 121–156.

[144] Lange, J.M.A., Paul, D.A., Hinsman, H.G., deWolf, F., van den Berg, H., Coutinho, R.A., Danner, S.A., van der Noordaa, J., Goudsmit, H. : Persistent HIV antigenaemia and decline of HIV core antibodies associated with transmission of AIDS. Br. Med. J. **293** (1986), 1459–1462.

[145] LaSalle, J., Lefschetz, S. : Stability by Lyapunov's Direct Method, Academic Press, New York, 1961.

[146] Leung, A. W. : Limiting behavior for a prey-predator model with diffusion and crowding effects. J. Math. Biology, **6** (1978), 87–93.

[147] Leung, A. W. : Systems of Nonlinear Partial Differential Equations. Applications to Biology and Engineering. Kluwer Academic Publishers, Dordrecht, 1989.

[148] Levin, S.A. : Population models and community structure in heterogeneous environments. In : Levin, S.A. (ed.) Mathematical Association of America Study in Mathematical Biology. Vol. II : Populations and Communities. Math. Assoc. Amer., Washington, 1978, pp. 439–476.

[149] Levin, S.A. : Coevolution. In : Freedman, H.I. and Strobeck, C. (eds.) Population Biology. Lecture Notes in Biomathematics, Vol.52, Springer-Verlag, Heidelberg, 1983, pp. 328-324.

[150] Levin, S.A., Castillo-Chavez, C. : Topics in evolutionary theory. In : Mathematical and Statistical Developments of Evolutionary Theory. NATO Advanced Study Institute, Montreal, Canada, 1989.

[151] Levin, S.A., Pimentel, D. : Selection of intermediate rates of increase in parasite-host systems. Am. Nat. **117** (1981), 308–315.

[152] Levin, S.A., Segel, L.A. : Pattern generation in space and aspect. SIAM Rev. **27** (1985), 45–67.

[153] Lin, X. : Qualitative analysis of an HIV transmission model. Math. Biosci. **104** (1991), 111–134.

[154] Lin, X. : On the uniqueness of endemic equilibria of an HIV/AIDS transmission model for a heterogeneous population. J. Math. Biol. **29** (1991), 779–790.

[155] Liu, W.M., Hethcote, H.M., Levin, S.A. : Dynamical behavior of epidemiological models with nonlinear incidence rates. J. Math. Biol. **25** (1987), 359–380.

[156] Liu, W.M., Levin, S.A., Iwasa, Y. : Influence of nonlinear incidence rate upon the behavior of SIRS epidemiological models. J. Math. Biol. **23** (1986), 187–204.

[157] Longini, I.M.Jr., Scott Clark, W., Haber, M., Horsburgh, R.Jr. : The stages of HIV infection : waiting times and infection transmission probabilities. In : Castillo-Chavez, C. (ed.) Mathematical and Statistical Approaches to AIDS Epidemiology. (Lecture Notes in Biomathematics, vol. 83.) Springer-Verlag, Berlin, Heidelberg, 1989.

[158] Macdonald, G. : The Epidemiology and Control of Malaria. Oxford Univ. Press, Oxford, 1957.

[159] Macdonald, G. : The dynamics of helminthic infections, with special reference to schistosomes. Trans. Roy. Soc. Trop. Med. Hyg. **59** (1965), 489–504.

[160] Marek, I. : Frobenius theory of positive operators: comparison theorems and applications. SIAM J. Appl. Math. **19** (1970),607-628.

[161] Martin, R.H. Jr. : Asymptotic stability and critical points for nonlinear quasimonotone parabolic systems. J. Diff. Eqns. **30** (1978), 391- 423.

[162] Matano, H. : Asymptotic behavior and stability of solutions of semilinear diffusion equations. Publ. RIMS, Kyoto Univ. **15** (1979), 401- 454.

[163] May, R.M. : Population biology of microparasitic infections. In : Hallam T.G. and Levin S.A. (eds.) Mathematical Ecology. Springer-Verlag, Berlin, Heidelberg, 1986, pp. 405–442.

[164] May, R.M., Anderson, R.M. : Population biology of infectious diseases II. Nature **280** (1979), 455–461.

[165] May, R.M., Anderson, R.M., Mc Lean , A.R. : Possible demographic consequences of HIV/AIDS epidemics I. Assuming HIV infection always leads to AIDS. Math. Biosci. **90** (1988), 475–505.

[166] May, R.M., Anderson, R.M., Mc Lean, A.R. : Possible demographic consequences of HIV/AIDS epidemics II. Assuming HIV infection does not necessarily lead to AIDS. In : Castillo-Chavez, C., Levin, S.A. and Schoemaker, C.A. (eds.) Mathematical Approaches to Problems in Resource Management and Epidemiology. (Lecture Notes in Biomathematics vol. 81.) Springer-Verlag, Berlin, Heidelberg, 1989.

[167] McKendrick A. : Applications of mathematics to medical problems. Proc. Edinburgh Math. Soc. **44** (1926), 98–130.

[168] Metz, J.A.J., Diekmann, O. , Eds. : The Dynamics of Physiologically Structured Populations. Lecture Notes in Biomathematics, vol. 68. Springer-Verlag, Berlin, Heidelberg, 1986.

[169] Mora, X. : Semilinear parabolic problems define semiflows on C^k spaces. Trans. Am. Math. Soc. **278** (1983),21-55.

[170] Murray, J.D. : Nonlinear Differential Equation Models in Biology. Clarendon Press, Oxford, 1977.

[171] Murray, J.D. : Mathematical Biology. Springer-Verlag, Berlin, Heidelberg, 1989.

[172] Nåsell, I. : Mating models for schistosomes. J. Math. Biol. **6** (1978), 21–35.

[173] Nåsell, I. : Hybrid Models of Tropical Infections. Lecture Notes in Biomathematics, vol. 59. Springer-Verlag, Berlin, Heidelberg, 1985.

[174] Nåsell, I., Hirsch, W.M. : A mathematical model of some helminthic infections. Comm. Pure and Appl. Math. **25** (1972), 459–477.

[175] Nåsell, I., Hirsch, W.M. : The transmission dynamics of schistosomiasis. Comm. Pure and Appl. Math. **26** (1973), 395–453.

[176] Nold, A. : Heterogeneity in disease-transmission modeling. Math. Biosci. **52** (1980), 227–240.

[177] Okubo, A. : Diffusion and Ecological Problems : Mathematical Models. Springer-Verlag, Berlin, Heidelberg, 1980.

[178] Pao, C.V. : On nonlinear reaction-diffusion systems. J. Math. Anal. Appl. **87** (1982), 165–198.

[179] Pazy, A. : Semigroups of Linear Operators and Applications to Partial Differential Equations. Springer-Verlag, Berlin, Heidelberg, 1983.

[180] Piccinini, L.C., Stampacchia, G., Vidossich, G. : Differential Equations in $\mathbb{R}^n$. Springer-Verlag, Berlin, Heidelberg, 1984.

[181] Protter, M., Weinberger, H. : Maximum Principles in Differential Equations. Prentice-Hall, Englewood Cliffs, N.J., 1967.

[182] Pugliese, A. : Population models for diseases with no recovery. J. Math. Biol. **28** (1990), 65–82.

[183] Pugliese, A. : An SEI epidemic model with varying population size. Preprint U.T.M. **311**, Trento, 1990.

[184] Pugliese, A. : Stationary solutions of a multigroup model for AIDS with distributed incubation and variable infectiousness. Preprint U.T.M. **357**, Trento, 1991.

[185] Rauch, J., Smoller, J. : Qualitative theory of the FitzHugh- Nagumo equations. Adv. in Math. **27** (1978), 12–44.

[186] Redheffer, R. : Volterra Multipliers II. SIAM J. Alg. Discr. Math. **6** (1985).

[187] Redheffer, R.M., Walter, W. : On parabolic systems of the Volterra predator-prey type. Nonlinear Analysis, T.M.A. **7** (1983), 333– 347.

[188] Redheffer, R.M., Walter, W. : Solution of the stability problem for a class of generalized Volterra prey-predator systems. J. Diff. Equations **52** (1984), 245–263.

[189] Redheffer, R.M., Zhou, Z. : Global asymptotic stability for a class of many-variable Volterra prey-predator systems. Nonlinear Analysis, T.M.A. **5** (1981), 1309–1329.

[190] Redheffer, R.M., Zhou, Z. : A class of matrices connected with Volterra prey-predator equations. SIAM J. Alg. Discr. Math. **3** (1982), 122– 134.

[191] Ross, R. : The Prevention of Malaria. Murray, London, 1911.

[192] Rothe, F. : Convergence to the equilibrium state in the Volterra-Lotka diffusion equations. J. Math. Biol. **3** (1976), 319–324.

[193] Rushton, S., Mautner, A.J. : The deterministic model of a simple epidemic for more than one community. Biometrika **42** (1955), 126– 132.

[194] Salahuddin, S.Z., Markham, P.D., Redfield, R.R., Essex, M., Groopman, J.E., Sarngadharan, M.G., McLane, M.F., Sliski, A., Gallo, R.C. : HTLV-III in symptom-free seronegative persons. Lancet **2** (1984), 1418–1420.

[195] Sattenspiel, L., Simon, C.P. : The spread and persistence of infectious diseases in structured populations. Math. Biosci. **90** (1988), 341–366.

[196] Sattinger, D.H. : Monotone methods in nonlinear elliptic and parabolic boundary value problems. Indiana Math. J. **21** (1972), 979–1000.

[197] Schaefer, H.H. : Banach Lattices and Positive Operators. Springer- Verlag, Berlin, Heidelberg, 1974.

[198] Schenzle, D. : An age structural model of pre- and post-vaccination measles transmission. IMA J. Math. Appl. Biol. Med. **1** (1984), 169–191.

[199] Selgrade, J.F. : Mathematical analysis of a cellular control process with positive feedback. SIAM J. Appl. Math. **36** (1979), 219–229.

[200] Severo, N. C. : Generalizations of some stochastic epidemic models. Math. Biosci. **4** (1969), 395–402.

[201] Skellam, J.G. : Random dispersal in theoretical populations. Biometrika **38** (1951), 196–218.

[202] Smith, H.L. : Cooperative systems of differential equations with concave nonlinearities. J. Nonlin. Anal. T.M.A. **10** (1986), 1037–1052.

[203] Smith, H.L. : Systems of ordinary differential equations which generate an order preserving flow. A survey of results. SIAM Rev. **30** (1988), 87–113.

[204] Smoller, J. : Shock Waves and Reaction-Diffusion Equations. Springer-Verlag, Berlin, Heidelberg, 1983.

[205] Solimano, F., Beretta, E. : Graph theoretical criteria for stability and boundedness of predator-prey systems. Bull. Math. Biol. **44** (1982), 579–585.

[206] Takeuchi, Y., Adachi, N., Tokumaru, H. : The stability of generalized Volterra equations. J. Math. Anal. **62** (1978), 453–473.

[207] Thieme, H.R. : A model for the spatial spread of an epidemic. J. Math. Biol. **4** (1977), 337–351.

[208] Thieme, H.R. : Global asymptotic stability in epidemic models. In : Knobloch, H.W. and Schmitt, K. (eds.) Equadiff. (Lecture Notes in Mathematics, vol. 1017.) Springer-Verlag, Berlin, Heidelberg, 1983, pp. 608–615.

[209] Thieme, H.R. : Local stability in epidemic models for heterogeneous populations. In : Capasso, V., Grosso, E. and Paveri-Fontana, S.L. (eds.) Mathematics in Biology and Medicine. (Lecture Notes in Biomathematics, vol. 57.) Springer- Verlag, Berlin, Heidelberg, 1985, pp. 185–189.

[210] Verhulst, F. : Nonlinear Differential Equations and Dynamical Systems, Springer-Verlag, Heidelberg, 1990.

[211] Vogel, T. : Dynamique théorique et hérédité. Rend. Mat. Univ. Politec. Torino **21** (1961), 87–98.

[212] Volterra, V. : Sui tentativi di applicazione delle Matematiche alle Scienze Biologiche e Sociali. Giornale degli Economisti, Serie II, **23** (1901). Reprinted in : Volterra, V. : Saggi Scientifici. Zanichelli, Bologna, 1990.

[213] Volterra, V. : Leçons sur la Théorie Mathématique de la Lutte pour la Vie, Gauthier-Villars, Paris, 1931.

[214] von Foerster, H. : Some remarks on changing populations. In : Stohlman F. (ed.) The Kinetics of Cellular Proliferation. Grune and Stratton, New York, 1959, pp. 382–407.

[215] Walker, J.A. : Dynamical Systems and Evolution Equations. Theory and Applications. Plenum Press, New York, 1980.

[216] Waltman, P.: A Second Course in Elementary Differential Equations. Academic Press, Orlando, Fla., 1986.

[217] Wang, F.J.S. : Asymptotic behavior of some deterministic epidemic models. SIAM J. Math. Anal. **9** (1978), 529–534.

[218] Watson, R.K. : On an epidemic in a stratified population. J. Appl. Prob. **9** (1972), 659–666.

[219] Webb, G.F. : Compactness of bounded trajectories of dynamical systems in infinite dimensional spaces. Proc. Roy. Soc. Edinburgh, Sect. A, **84** (1979), 19–33.

[220] Webb, G.F. : Theory of Nonlinear Age-Dependent Population Dynamics. Marcel Dekker, New York, 1985.

[221] Williams, S., Chow, P.L. : Nonlinear reaction-diffusion models for interacting populations. J. Math. Anal. Appl. **62** (1978), 157–169.

[222] Wilson, E.B., Worcester, J. : The law of mass action in epidemiology.
- Part I, Proc. Nat. Acad. Sci., **31** (1945), 24–34,
- Part II, Proc. Nat. Acad. Sci., **31** (1945), 109–116.

[223] Wilson, H.K. : Ordinary Differential Equations. Addison-Wesley, Reading, Mass., 1971.

[224] Yang, G. : Contagion in stochastic models for epidemics. Annals of Math. Statistics **39** (1968), 1863–1889.

Notation

$\mathbb{N}$	$:= \{0, 1, 2, \cdots\}$ is the set of all natural numbers
$\mathbb{N}^*$	$:= \mathbb{N} - \{0\}$
$\mathbb{R}$	is the set of all real numbers
$\mathbf{C}$	is the set of all complex numbers
$\mathbb{R}_+$	$:= [0, +\infty)$ is the set of all nonnegative real numbers
$\mathbb{R}_+^*$	$:= \mathbb{R}_+ - \{0\} = (0, +\infty)$ is the set of all positive real numbers
$\mathbb{R}^n$	$:= \mathbb{R} \times \cdots \times \mathbb{R}$ (n times) is the n-dimensional Euclidean space
$\mathbb{K}$	$:= \mathbb{R}_+^n = \mathbb{R}_+ \times \cdots \times \mathbb{R}_+$ is the positive cone of the n-dimensional Euclidean space
$\mathbb{R}_+^{n*}$	$:= \mathbb{R}_+^* \times \cdots \times \mathbb{R}_+^*$ (n times) $= \overset{\circ}{\mathbb{K}}$
$\Vert \cdot \Vert$	denotes in general the Euclidean norm in $\mathbb{R}^n$, but it can also denote the norm in an arbitrary normed vector space (depending upon the context)
$B_\rho(z_0)$	$:= \{z \in E \mid \Vert z - z_0 \Vert < \rho\}$ denotes the open ball with center $z_0 \in E$ and radius $\rho > 0$ in a normed vector space E
A^T	denotes the transpose of the matrix A

Subject Index